The R.A.M.S. Library of Alchemy

Volume 4

The Golden Work of Hermes Trismegistus

Hermes Unveiled
by Cyliani

R.A.M.S. Publishing Company

The Golden Work of Hermes Trismegistus
(Tractatus Aureus)

Hermes Unveiled
by Cyliani

Produced by

Restorers of Alchemical Manuscripts Society

R.A.M.S. Publishing Company

HERMES

Hermetis Trismegisti Tractatus Aureus

Redrawn from an original manuscript dated 1577.

THE LEAVES OF HERMES' SACRED TREE.
In his Key to Alchemy, Samuel Norton divides into fourteen parts the processes or states through which the alchemical substances pass from the time they are first placed in the test tube until ready as medicines for plants, minerals, or men:

Printed for *J. Harris* and *T. Hawkins*   1692

The Golden Work of Hermes Trismegistus, Translated out of Hebrew into Arabick, then into Greek, afterwards into Latin; and now done out of Latin into English, Claused and largely Commented upon by:

WILLIAM SALMON

PROFESSOR OF PHYSICK

5

R.A.M.S. Publishing Company
117 Rutherford Lane
Stuarts Draft VA 24477

First Edition 2015

ISBN-13 **978-1508598213**
ISBN-10 **1508598215**

Printed in the United States of America

Dedicated to Hans W. Nintzel,
American Alchemist
and
Founder of the
Restorers of Alchemical Manuscripts Society
(R.A.M.S.)

Disclaimer

Liability: The publisher does not warrant or assume any legal liability or responsibility for the accuracy, completeness, or usefulness of any information, apparatus, product, or process disclosed. The publisher makes no representation as to the accuracy or completeness of the contents of this book and specifically disclaims any implied warranty of merchantability or fitness for a particular purpose. No warranty may be created or extended by written sales materials or sales representatives. You should obtain professional consultation where appropriate. The publisher shall not be liable for any loss of profit or other commercial or personal damages, including but not limited to special, incidental, consequential, or other damages.

Table of Contents

Introduction

Philip N. Wheeler

Hermes Trismegistus, or thrice-greatest Hermes, may have been the author of a number of Alchemical texts. Some speculate that he was a mythical creation from a combination of the Greek god Hermes and the Egyptian god Thoth. The writings attributed to Hermes had a decisive effect on the Renaissance. Perhaps the most famous work attributed to Hermes is "The Emerald Tablet" that includes the passage:

> That that which is Above is like that which is Below and that which is Below is like that which is Above, to accomplish the Miracle of Unity.

Francis Barrett states in his book *The Magus*, "HERMES Trismegistus, (who was the author of the divine Pymander and some other books,) lived some time before Moses. He received the name of Trismegistus, or Mercurius ter Maximus, i. e. thrice greatest Intelligencer, because he was the first intelligencer who communicated celestial and divine knowledge to mankind by writing."

Hans Nintzel considered *The Golden Work* an essential source for the student of Alchemy.

Also included in the Volume is the valuable work, "Hermes Unveiled" by Cyliani. This was offered as a separate manuscript by Hans, but in itself it is too short to be printed as a standalone book.

TRACTATUS AUREUS

CHAPTER I

THE PREFACE EXPLICATING, IN PART, THE PRIMA MATERIA.

I. Hermes. Even Hermes himself saith, I have not in a very long Age, ceased to try Experiments, nor have I spared any Labour of mind: But I obtained the knowledge of this Art, by the Inspiration of the Living God only, who esteeming me his Servant worthy, did reveal and open the Secret to me.

Salmon. There are three things which are certainly most necessary to the attainment of this knowledge. 1. An Unwearied Study. 2. A Continued Experience. 3. And the Devine Blessing going along with all. Without these, it is not probable any Man can attain the knowledge of this Secret. There must be a diligent Study, and a ferious Meditation in the Soul, concerning this thing: Then these things thus meditated on, must, by experience, be brought to ocular demonstration; nor, if you miss many times, must you be weary with trying. Lastly, you must all along attend the Blessing of God for his assistance:

'Tis that Eternal Spirit of God which goes through,

and pierces all things, which generates, and preserves that which is generated: His Spirit of heat decocts, and coagulates that which is thin, rarifies that which is too thick, warms the cold; and raises up to life that which has been dead and buried.

II. Hermes. Who has given to, or bestowed upon rational Creatures, the power and faculties of truly judging and determining, not forsaking any, so as to give them an occasion to cease searching after the Truth.

Salmon. 'Tis true, that Adam before the Fall was adorned with the fulness of light and knowledge above all other Creatures, shining like Sol among the Stars; but after his Fall, that prime perfection was much eclipsed, and he was drove out of the Garden, into a Wilderness among the Beasts which perish; yet not without a promise of Restauration, and remission of his Transgression, by one Eternal Sacrifice, through the diffluence and power of Whole Spirit, Man is put into a possibility of attaining a measure of the true and perfect knowledge and understanding even in this life.

III. Hermes. For my part, I had never discovered any

thing of this matter, nor revealed it to any one, had not the fear of the Judgments of God, or the hazard of the Damnation of my Soul for such a Concealment prevailed with me. It is a debt I am willing to pay to the Just; as the Father of the Just has liberally bestowed it upon me.

Salmon. That is, revealed them so as that the Sons of Art might understand them, not to the Profane and Unworthy, and Scoffers: For the Oracle of Truth himself has long since told us, It is not fit to give the Childrens Bread to Dogs; though they may eat of the Crumbs which fall from the Masters Table. Some Men the Scriptures of Truth have compared to Dogs, yea, Greedy Dogs, Wolves, Foxes, These can never come to sit at the Table, and feed of the Divine repast; 'Tis a Transgression against the Law of Nature, which is the Law of God, which deserves the Divine Vengeance for a punishment: And such indeed is the revealing of forbidden Secrets to such to whom they do not belong. And saith Raimand Lully, Thou shalt reserve and keep that Secret, which is proper only to God to reveal, and thou dost justly conceal those things, whose revelation belongs to his Honour; otherwise thou shalt be condemned in the Great day, as a Traytor to the Majesty of God, nor shall thy Treason be forgiven thee.

IV. Hermes. Now understand, O ye Children of Wisdom, that the knowledge of the four Elements of the Ancient Philosophers, was not Corporally, nor Imprudently sought into: Which is through patience to be attained, according to their kind, which through their own operation are hidden or obscured. You can do nothing, except the matter be compounded, because it cannot be perfected, unless first the various Colours are throughly accomplished.

Salmon. Hermes now begins to give a description of the Great Work, which he calls the knowledge of the Elements, but not of those Elements which are foolishly discoursed of in the Schools of the Peripateticks: They speak of an Element to be Corpus Simplex, but our Hermes saith, They are not to be understood Corporaliter. Spiritualiter Sapienter, that is, Spiritually and Wisely. Thus the Principles of Art are said to be four Elements, Earth, Water, Air, Fire, as Hermes indigitates, but what these are in a Spiritual sense, the Peripatelick knows not, which the same Hermes interprets in another place, the Soul, Spirit, and Body; and which Paracelsus calls Salt, Sulphur, and Mercury. Others make but two, as the Agent and Patient; Masculine and Feminine; Sulphur and Mercury: Others but one only, viz. The Aqua Philosophica. There are many other Names by which this

16

Matter is called, but the Subject, or Prima materia, is one only: because it is, as it were, the Cardinal hinge upon which the Philosophers explicate to be their Mercury, which is the beginning, the middle, and the end of the Work, and without which, whoever labours, labours in vain; and yet it will do nothing without it be compounded, because it cannot be perfected without its colours are throughly accomplished: The Body and the Soul; or the Salt and the Sulphur, cannot be united in their most minute parts, without the help of the Spirit which is Mercury. Luna and Sol cannot procreate without the help of Mercury, which extracts, the Semen from both the Bodies, and in the center of the Earth, as its proper Vessel, digests and perfects it. Therefore Mercury does nothing of its self, except something be added to it by which it may be mortified.

V. *Hermes. Know then, that the Division which was made upon the water by the Ancient Philosophers, is that which separates it, or converts it into four other substances; one into two, and three to one; the third part of which is color, or has tincture, to wit, the coagulating humour or moisture, but the second and third Waters are the Weights of the Wise.*

Salmon. This Water to be divided, is the same with the

four Elements before spoken of, viz. The Aqua Philosophica: This must be divided into four parts, viz. The one part into two; adding three parts to one; from whence arises seven parts: He divides the differences of the Colors into two threes, that is, into three Red Spirits, and three White, which three Spirits have their rise from the one Aqua Philosophica, and are resolved into the same again.

VI. Hermes. Take of the humidity or moisture, an ounce and half; of the Southern Redness, viz. Anima Solis, a fourth part, i.e. half an ounce, of the Citrine Seyre in like manner half an ounce: of Auripigment half an ounce, which are eight; that is three ounces: Now understand that the Vine of the Wise Men, or Tree of the Philosophers is extracted or drawn forth in three, but the Wine thereof is not perfected till at length thirty be compleated.

Salmon. He Essays to explicate the proportions of the Philosophick Ingredients, under various Names; for that which he calls the Humidity, the Southern Redness, Anima solis, Seyre Citrinum, Auripigment, the Vine of the Philosophers, and their Wine, have no other signification, but that the Aqua Mercurii should be seven times distilled, which after the eighth Distillation, the Compositum, by the force of the fire, is converted into Ashes, or a most subtil pouder, which by reason of its purity and perfection resists the fire: neither wonder that eight parts and three ounces are equivalent, for by the former Section, the one part is divided into two, to each of them, there is added three parts, which are the true and Philosophick Proportions, called by Hermes, the Weights of the Wise.

VII. Hermes. Understand then the operation. The Decoction doth diminish the matter, but the Tincture does augment it: Because Luna in 15 days is diminished (in the Heaven) and in the third operation (viz. after the Conjunction with Sol) it is augmented. This is then the beginning, and the end.

Salmon. Here Hermes eludicates the Philosophick Work by a most familiar Example of the Phases of Luna; and so it is, the Mineral Process in this Philosophick Work, exactly answering to that Parallel in Heaven. Some divide the Operation of the Stone into two parts, viz. the former and the latter. The former Hermes explicates by the notion of Decoction, which does diminish the matter, dissolves it, as it were destroys it; but being thus Dissolved and Corrupted, it is through Regeneration (by the Medium of perfection) restored again. This done, then follows the latter part of the Operation, by means of which the Virtue and Power of the Stone is made wonderful, brought to its highest perfection, and multiplied (as it were) in infinitum. In these few words of Hermes, are comprehended the whole Work, and in them it is plainly laid open from the beginning to the end. In a word, it is like the Husbandman Sowing his Seed in the Ground, which must first Die, be Corrupted and Putrefied, before it can be possest of a new Life, by which it must arise and yield its Hundred Fold Increase: the first Life, the first Birth, the first Body, must Die, and give place to the second.

CHAPTER II

THE FIRST EXPOSITION OF THE MATTER.

I. Hermes. Behold, I have Exposed to you that which was hidden, and the work is both with you and for you: that which is within, is quickly taken out, and is Permanent or fixt; and you may have it either in the Earth, or in the Sea.

Salmon. This secret Work, commends itself to its Children; and the series of the Operation demonstrate, that the Regenerating Spirit is within the matter, but adhears to it invisibly. In Elementary and Gross Bodies, it is not manifest, except they be reduced into their first Essential Nature or Being; for so this Spirit of Regeneration which is the Seed of the Promise, the Heaven of the Philosophers, out shining the Glory of the Stars, is brought forth to View. That which is Sown is not quickned except it Die; it is Sown in Corruption, it Rises in Incorruption, it is Sown in Dishonour, it is Raised in Glory. The Sea is the Aqua Philosophica, which entring into, and Opening the Terra Philosophica, brings forth the old bearing Vine of the Philosophers.

II. Hermes. Keep therefore your Argent Vive, which is

prepared in the innermost Chamber of the Bridegroom, in which it is Coagulated; for that is the Argent Vive itself, which is spoken of the remaining Earth.

Salmon. Argent Vive is indeed, the Prima Materia of the Philosophick Work, but (say the Philosophers) beware that you use not the Vulgar Argent Vive, or Quick-Silver; for if you do, you will be deceived. Our Silver is not Vulgar, for that is Dead, and unfit for Our Work; you must have that which is Living, which is rightly Prepared by Art for the perfection of Nature. Our Mercury is Philosophick, Fiery, Vital, Running, which may be mixed with all the other Metals, and separated again from them. It is prepared in the innermost Chamber, there it is Coagulated: Now, where Metals grow, there they must be found: If you have found this Argent Vive, the residence of the Philosophick Earth, keep it safely, for it is worthy: If you have brought your Argent Vive to Ashes, or Burnt it by the Power of the Fire, you have an incompatable Treasure, a thing much more Pretious than Gold. This is that which Generates the Stone, and it is Born of it, it is the whole Secret, which Converts all other Metalline Bodies into Sol and Luna, making Hard Soft; and the Soft Hard, putting Tincture and Fixity upon them.

III. Hermes. He therefore that now hears my Words, let him search into, and inquire, from them; it is not for the justification of the Work of any Evil Doer, but to give to every good Man a Reward, that I have laid Open or Discovered all things which were bid, relating to this Science; and Disclosed and made Plain and Open to you the greatest of Secrets, even the Intellectual knowledge.

Salmon. The Philosophers ever Discourse in Parables and Figures; not is it fit that all things should be revealed to every Body; the matter is to be enquired after, and diligently Searched into; without Labour and Pains, nothing is to be obtained; but Wisdom enters not into profane Souls, nor dwells in a Body subject to sin, as the Wise Man affirms. And altho' Hermes has spoken in this Book many things concerning this most noble Arcanum, and has over past nothing, yet he has not spoken so plainly as that every profane and unworthy Person may understand it, but has left the Mystery to be unfolded by the Sons of Wisdom.

IV. Hermes. Know therefore ye Children of Wisdom, and ye seekers after the Fame thereof, that the Vulture standing upon the Mountain, cries out with a great

Voice; saying, I am the White of the Black, and the Yellow of the White, and the Citrine of the Yellow, and behold I speak the very Truth.

Salmon. The Mountain upon which the Vulture stands, is a fit Vessel placed in a well Built Furnace, encompassed with a Wall of Fire; at the foot of which Mountain is a watchful Dragon, who is full of Eyes, and can see before him and behind him, who is Vigilant and Careful in keeping the Entrance or Passage into the Mountain, left the unworthy should Ascend to the height thereof, where is hid the Secret Stone of the Philosophers: It is impossible for any to enter here, unless the Dragon be laid a Sleep; Hoc opus, bic Labor, to find out the means how this is to be done, how this Beast is to be circumvented, that we may obtain this so desirable Treasure is the Work of the Philosopher. Three things are commended for this purpose, first Crude Argent Vive made into Pills, and Gilded with Gold. Second, a Sulphur of Mars extracted with Sol. Third, The water of the Philosophers. These things being rightly given, will so lay him a Sleep, that Night and Day you may continually have Egress and Regress. Being once entred, and Ascended the Mountain, the Vulture or Crow will shew you the way where the Colors appear.

1. Black which is the beginning of the Art.
2. White which is the middle.
3. Red which is the end of the whole Work.

V. Hermes. Now the chief principle of Art is the Crow, which in the Blackness of the Night, and Clearness of the Day, flys without Wings, From the bitterness existing in the Throat, the Tincture or Tinging matter is taken: But the Red goes forth of its Body, and a meer Water is taken from its back parts.

Salmon. The Vulture and the Crow, are both but one thing, but in differing States, it is the Vulture while it is Active and devouring; and the Crow when it lies in a more passive Nature. The Vulture is the Mercury of the Philosophers prepared by the help of Vulgar Argent Vive: And the Crow is the Infancy of the Work, wherein the said Philosophick Mercury is United with its Solar Ferment. The blackness of the Night is the Putrefaction thereof, and the clearness of the Day, its Resurrection into a State of Purity. It flies without Wings, being Born or carried by the fixt Nature; and the bitterness in the Throat, is the Death of the first Life, whence is Educed the Soul, which is the Red and Living Tincture taken from the Body: And the Water is the Viscous Humidity, made of the Philosophers Argent Vive, which radically dissolves all Metals, and reduces them into their first Ens or Water; and also reduces common Quick—Silver into the fame, by a Simple Imbibition, for ever.

VI. Hermes. Understand and accept of this gift of God, which is hidden from Ignorant and Foolish Men. This hidden Secret which is the Venerable Stone, splendid in Color, a sublime Spirit, an Open Sea, is hid in the Caverns of the Metals: Behold I have exposed it to you; and give thanks to the Almighty God, who teaches you this knowledge; If you be grateful, he will return you the Tribute of your Love.

Salmon. Fools, and unlearned, are excluded from the knowledge of this Mystery, viz. Such as are unacquainted with the gift of God; which is a measure of his Holy Spirit. He calls it a Stone, yet says, it is a Spirit; for was it not a Spirit, it could not Penetrate and Tinge other Bodies by an absolute Unity and Conjunction: Bodies and Matter cannot do this, the most that they can do is but to touch one another in their Superficies; for all matter is Dead, and no Dead thing can penetrate into the property of another, but only (at most) lie side by side with it. And to make the matter the more sensible to your understanding, he compares it to an open Sea, for that this Spirit peirces Bodies, and is joyned to them, even as Water

26

is joyned to Water, or as the Salt Body thereof is joyned with its Aqueous parts. It is hidden in the Caverns of the Metals, that is, if you seek for it in any thing that is not Metalline, you stumble at the Threshold.

VII. Hermes. You must put the matter into a moist fire, and make it to Boil, which Augments the Heat of the Humour or Matter, and destroys the Dryness of the incombustible Sulphur; continue Boiling till the Radix may appear then Extract the Redness and the light parts, till only about a third remains.

Salmon. There are said to be three Species of Decoction.

1. An external Fiery heat in Humido, and is called Elixation.
2. An external heat in Sicco, which is called Assation.
3. An internal natural heat in Humido, called by the Greeks, i.e. Maturity, or the Ripening and perfecting heat.

Now which of these it is, that Hermes speaks of is the question. The first, and the third differ in this. The first is an external Fiery Heat. The latter an

Internal Natural Heat. In my Opinion both are to be admitted. The Natural Heat Internal, is the Cause of Generation and without that, the External Heat can do nothing: Hence we conclude the Heat to be two—fold. 1. External to excite. 2. Internal to perfect, both which ought to be made in humido: for all Generation is naturally made in Calido Humido, in a moist Heat, which Hermes calls Ignem Humidum: as if he should say the Fire is two-fold, which you must use, viz. External and Internal. He seems to make his Coction double, 1. In the time of Augmentation. 2. In the Ultimate perfection or Maturity, and so long this Fire is to be continued, till the Radix does appear, i.e. the Seed of Metals. The same method that Nature takes in Generating Herbs and Plants, she takes in Generating Metals, whose Seed is extracted by the help of Art, which Seed is only and truly the Philosophers Mercury, in which all the Metals are resolved into their first principles, and in which is imprest the Character or Power of Transmutation. They all err who think to reduce Metals only into Crude Mercury, and not into their Radix, as Hermes speaks, viz. into their Seeds, which is the first Matter living in Metals: and from thence Nature evergoes forward, never back-ware till she comes to perfection.

VIII. *Hermes. For this Cause-sake, the Philosophers are said to be Envions or Obscure, not for that they*

Grudged the thing to the honest or just Man, to the Religious or Wise; or to the Legitimate Sons of Art? but to the Ignorant, the Vitious, the Dishonest: lest evil Persons should be made powerful to perpetrate sinful things: for such a fault the Philosophers must render an account to God. Evil Men are not worthy of this Wisdom.

Salmon. It appears that neither Hermes, nor any of the other Philosophers did Envy or Grutch the true knowledge of the matter to the Pious, Just, and good Man, but only to the Profane and Wicked, they did not think it fit to give the Childrens Bread to Dogs, for which Cause-sake, they always keep the Prima Materia Secret, and lest it as a Legacy to the Legitimate Sons of Art; but the manner and way of working it, through all its various Operations, they have faithfully and plainly declared to the least Iota, or Tittle.

CHAPTER III

THE NAMES AND FIRST OPERATION EXPLICATED.

I. Hermes. Now this Matter I call by the Name of the Stone; the Feminine of the Magnesia, the Hen, the White Spittle, or Froth, the Volatile Milk, the Incombustible Ashes; so that is might be hidden from the simple and unwise, who want understanding, honesty, and goodness: which notwithstanding they signified it to the Wise and Prudent by one only Name, which is the Stone of the Wise, or the Philosophers Stone.

Salmon. There are various Names, by which the Philosophers call it, as Sol, Gold, Brass of the Philosophers, Magnesia, the pare Body, clear Ferment, Elixir, Masculine, Fixt Argent Vive, Incanbustible Sulphur, Red fixed Sulphur, the Rubin, Kibrick, Green Vitriol, the Greenness, Redness, burnt Brass, Red Earth, the Water of Sulphur, Aqua Mundi, Spittle of thna Shadow of the Sun, Eyes of Fishes, Sulphur, Sharp Wine, Urin, Light of Lights, Father of Minerals, Fruitful Thee, Living Spirit, Venom, moist strange Vinegar, White Gum, Everlasting Water, Aqua Vitae, a Woman, Man, Masculine, Feminine, a Vile thing, Azot, First Matter, Principium Mundi, and therefore Argent Vive, Mercury, Azot, Plessilunam, Hypostasis, White

Lead, Red Lead, Water, the Crow, Iron, Silver, Lime, Jupiter, Vermilion, Whiteness, all signifie but one thing, Our Stone, but in diver times and degrees of Operation. So also, White Earth, White Sulphur, Ethel, Auripigmentum, Arsenick, Chaos, a Dragon, Serpent, Toad, Green-Lyon, Red-Lyon, Camelion, Quintessence, Virgins Milk, Radical Humidity, Unctuous Moisture, Sperm, Sal Armoniac, Hair, Urine, Antimony, Philosophers Lead, Salt, a Bird, Microcosmus, Cinnaber, do all signifie but one and the same thing.

II. Hermes. Conserve therefore in this the Sea, the Fire, and the heavenly Bird, even to the last moment of its Exit. But I deprecate, or wish a Curse from our Benefactor, the great and Living God, even to all the Sons of the Philosophers, to whom it shall please God to give of the Bountifulness of his Goodness, if they shall undervalue, or divulge the Name and Power thereof, so any Foolish or Ignorant Person, or any Man unfit for the knowledge of this Secret.

Salmon. He teaches here, that in the matter of the Stone, is to be Conserved the Sea, the Fire, and the Heavenly Bird, to the Perfection or Consummation of the Word; by the Sea, is understood the Humidity of the Mercury, for that no Generation can be made in a dry, but in a humid matter. Therefore Mercury is to be

Conserved in a Liquid form, citra tamen sui Corruptionem, but without its Putrefaction; for that hard things or Bodies, as Raymund Lully saith, receive not the Heavenly Virtue, nor yield to the heavenly Influences. A Seal puts not its Print upon a hard Stone, but upon soft wax: so our matter, by being made soft and Rarified, is made fit to receive the influx of the superior Bodies, i.e. of Sol and Luna, and is made to obey the Government of the Sun. By the Fire and the Heavenly Bird, is understood the two-fold Fire, the External and the Internal, with both which it is to be conserved and nourished to the end of the Work.

III. Hermes. Whatever any Man has given to me, I have returned it again; nor have I been behind hand with any, or desisted to return an equal kindness; even in this Friendship and Unity consists the chief matter of this Operation.

Salmon. This not only demonstrates the Generous and Noble Spirit of our great Hermes; but also the relation, which the parts Composing this Magistery have one to another; for saith he, even in this Friendship and Unity consists the chief matter of this Operation.

IV.   Hermes. This is the concealed Stone of many Colors, which is Born and brought forth in one Color only: Understand this and conceal it.

Salmon. By the many Colors, here is understood the Black, White, and Red, of which we have spoken before: and tho' there may appear many other Colors in the course of the Operation, yet those three are the chief; of which, the one Color which for ever remains, is the Never-fading Red, than which, nothing can be more noble or perfect; this, if thou attainest to be an Adept, a true Son of Art, be sure to hide and conceal it, as here thou art admonished.

V.   Hermes. By means of it, (through the permission of the Omnipotent) the greatest Disease is Cured; and every Sorrow, Distress, Evil, and hurtful thing may be Evaded: and through the help thereof, you may come from Darkness to Light; from a Desert or Wilderness to a Habitation or Home; and from straightness and necessities, to a large and ample Fortune.

Salmon. This Our Tincture, Our Elixir, Cures not only all the Diseases of Metals, but all the cureable Diseases in Humane Bodies: It gives also, not only Health and long Life, but removes Poverty and Want,

and the attendant Evils of a narrow and pinching
Fortune. It is indeed the great preservative against
all the Afflictions, Sorrows and Miseries of Humane
kind, of what Nature and quality soever. It is Nectar
and Ambrosia, to all the Vital Powers, through the
Efficacy of which, Nature is made able to contend,
resist, and overcome all her adversaries.

CHAPTER IV

A CONTINUATION OF THE EXPLICATION OF THE FIRST OPERATION.

I. *Hermes. Now my Son, before all things, I admonish thee to fear God, in whom is the Blessing of your undertakings; and the Uniting and disposing of every thing which you Segregate, put together, or Design for this purpose.*

Salmon. This great Philosopher well knew, that the only way to attain to the Mystery, was to be acquainted with that Spirit which knew all things, yea the deep things of God; and to be acquainted with that Spirit, was to fear God, for so says the Holy Spirit it self. The fear of the Lord is the beginning of Wisdom, and the Knowledge of the Holy is understanding: And therefore our Hermes advises us, above all things to fear God, in whom is the Blessing of this undertaking. He shall not Err, who becomes acquainted with, and joyned to, that Spirit which is the Fountain of all Knowledge and Wisdom: For being United with that, you are Centred into the very Root from whence all Wisdom and Knowledge spring, and being Ingrafted into that Root, the true understanding will grow up in you, and fill you even as the Soul is filled with Life.

II. Hermes. Whatever I speak or write, consider it, and reason about it in your mind: I advise not them who are depraved in their Reason and Understanding, nor the Ignorant, or Insipid of Judgment. Lay hold of my instructions, and Meditate upon them and so fit your Mind and Understanding (to conceive what I say,) as if you your self were the Author of these things I write.

Salmon. He here speaks to such as fear God, not to be depraved in their understandings, (as all Profane and Wicked Men are) nor the Ignorant (who are unacquainted with the true Fountain of knowledge, which is the Spirit of the Living God, as he himself has instructed, Chap. 1, Sect. 1. aforegoing:) nor to the Insipid of Judgment, (who has not Pondered nor Meditated upon these things.) You must Enter with your Spirit and Soul into the Center of Nature, and there behold how all things are begun, continued, and perfected; but you must first enter into that Universal Spirit, which is the Former of all things, which pierces through and dwells in that Central Root; and by entering into that, it will, like as a Vehicle, carry you into the same Root, where all things are hidden, and reveal to you the most abscondite Mysteries, and shew you as in a Glass the whole work, and all the Operations of Nature.

III. Hermes. For to what Nature is hot, if it shall be made cold, it shall do no hurt or injury to it; so in like manner, he to whom Reason is become a guide, does shut against himself, the Door of Ignorance, lest he should be securely deceived.

Salmon. That is, if the Spirit and Soul, or hot Mercury and Sulphur be made more cold, by a Conjunction with the cold Body, you shall not do a miss, but proceed rightly on in the Work: and this you must apprehend by your Reason, and the Nature of the thing. He to whom (saith he) Reason i.e. the Spirit of knowledge is become a guide, does shut from himself the Door of Ignorance, i.e. open to himself the Door of knowledge, leading into the Mysteries of this Our Philosophick Work.

IV.   Hermes. Take (my Son) the Flying Bird, and Dround it Flying; then divide, separate, or cleanse it from its Filth, which keeps it in Death; expel it, and put it away from it, that it may be made Living, and answer thee, not by Flying in the Regions above, but truly by forbearing to Flie.

Salmon. In this our Art are two Principles which
spring from one Root, and which are the subject of our
Stone, viz. Argent Vive, and Sulphur, of which, the
one is Volatile and Superior, the other fixt and
below, from the Conjunction of which often repeated,
is made the true and Philosophical sublimation and
fixation. And that is the fixation when the Body
receives the Tinging Spirit, and takes from it its
Volatility; this is done by frequent Reiterations,
till a Calx of perpetual duration is produced, and
will remain for ever in the Fire. But in the very
beginning of this work the substance of the Stone,
which in it self is most fixed, by a Spirit not fixed
or Volatile, as Sea Water, acetum radicatum, and such
like, is to be made Volatile. And by this means it
will be more fit to be cleansed of its Filth, or Rust,
which in metals is a most certain sign of
Imperfection.

*V.   Hermes. If therefore you shall deliver it out of
its Imprisonment, or Cage, or Straights then
afterwards you shall Order and Govern it, according to
the number of days I shall note to you, according to
Reason; and then it shall be a Companion to thee, and
by it, thou shalt be made great and powerful.*

Salmon. That is, the fixed Body is to be lifted up by

sublimation, and to be so often repeated, till the Volatile is made fixed with it: But this is not to be done hastily, or all at once, but by little and little, and by degrees. Lest by too great a haste you break the Vessel, or come to some other hurt. God himself, made all things in Number, Weight, and Measure; that is, in due and just proportion; as well in respect of Time as Matter. If you proceed wisely in this Case, you will receive the fulness of your expectations.

VI. Hermes. Extract from the Sun Beams the Shadow, and the sordid Matter, by that which makes the Clouds hang over it, and Corrupts it, and keeps it from the Light, because by its Torture and Red Fiery Heat, or Redness it is Burned.

Salmon. The shadow always goes along with the Body, walking in the Sun. Now that a clearer Light may appear through the Body, without any shadow, the Body must be opened, made thin, and dissolved; which is the Patient, by the Spirit or Sun-Beams, which are the Agent, the living Fire, by whose Power it is brought to a Calx, and the Corruptible part is Burnt up and destroyed, or made fit to be separated.

VII. Hermes. Take this Redness Corrupted with the Water (which resembles the Matter, holding the Fire as in a live Coal) from it: As often as you take this Redness Corrupted in Water, away from it, so often you have the Redness Purified, then will it Associate it self, viz. become fixed, and Tinged, in which station it will rest for ever.

Salmon. That is Our Magnesia, which is sown in our Philosophick Earth, is to be Corrupted or Putrefied; and then to be Digested, Coagulated, Sublimated, Incerated, and Fixed. This Magnesia, or Redness is thus made pure by separation, and then it becomes Dissolved, Digested, Coagulated, Sublimed, Incerated and Fixed, and Tinged, being first lift up into the highest Heavens, and then Buried again in the deepest Earth, that therewith it may arise, and in the same have a Habitation, and be fixt for ever. The Water is the Spirit; the Redness is the Soul or Tincture; and the Earth is the Body. Now the Spirit is the Life of the Soul, as the Body is the Clothing or Habitation thereof: so that the Body is a substance, fixed, dry, and containing both the Spirit and the Soul. The Spirit Penetrates the Body; the Body fixes the Spirit, the Soul conjoyned, Tinges of its own Color, whether it is White or Red.

VIII. Hermes. Return the Coal, being extinct in its life, to the Water, in the thirty days I note to thee, so will you have a Crowned King, resting upon the Fountain or Well, but drawing it from the Auripigment, and wanting the Humour or Moisture: Now have I made the Hearts of the attentive, who hope in thee, glad, and their Eyes beholding thee, in the hope of that which thou possessest?

Salmon. The Life of the Coal is Fire, which being removed from it, is like a Dead Body; nor in a Coal only, but in all other things, it is Fire that excites or stirs up up the Life, comforts it, and conserves it: Yea, the Essence of Life is nothing else than Pure, Naked, Unmixed Fire: not that which is Corrupting and Elementary but that which is Subtil, Coelestial, and Generating all things. This in Metals is the Aqua Philosophica, Oleaginosa, and Sulphurea, and in this the Earth is to be raised up in the space of 30 days, which is a certain Number for an uncertain: By the Crowned King, is meant the perfection of the Tincture. The Well is the Fountain of the Philosophers, inexhaustible; out of which it draws the Auripigment or eternal Tincture, but wanting indeed its moisture, or Running—Water, which is for so long time to be Digested and Boiled with Fires.

IX. Hermes. Now the Water was first in the Air, then in the Earth: restore thou it then, to the superior places, through its own meanders or passages, and (not foolishly or indiscreetly) change or alter it: and then to the former Spirit gathered in its Redness, you must carefully and leisurly joyn it.

Salmon. Convert the Elements, and you shall have what you seek. The Earth which is Cold and Dry, agrees with the Water in one quality, which is cold. The Air which is Hot and Moist, participates with the Fire which is Hot and dry; and consequently the Earth with the Fire, because of its Driness. The Earth is the only true and first Element of the Stone, which by a Philosophical Calcination is to be Burnt up, and Rarified, afterwards to be Dissolved in a Moist place into a Ponderous Water: This by Sublimation is made more subtil and converted into Fire. This Oyl by a most strong Fire into Ashes, or a Red Rubicund Earth. Thus the Dragon devours his own Tail; and the Pelican with her own Blood, nourishes her Young ones. The Blood of the Pelican is this Red Spirit. Now nothing is joyned together with it, but that which before was separated from it. This Mixtion of the Elements is not Corporeal, but Spiritual; not with Hands done, but the work of the Metalline Archeus or Spirit, which you ought well to know, and then you will not longer from the Truth.

CHAPTER V

A DIALOGUE BETWEEN HERMES AND HIS SON.

I.   *Hermes. Know thou my Son, that the fat of our Earth is Sulphur; that Sulphur is Auripigment, Siretz, or Colcothar, of which Auripigment, Sulphurs, and such like, some are more vile or mean than others, in which there is a difference of diversity. Of this kind also is the Fat of Glewy substances; to wit, of Hair, Nails, Hoofs, and Sulphur itself; Oyl of Peter and the Brain or Marrow, which is Auripigment. Of the same kind also is the Cats or Lyons Claw, which is Sirezt: The Fat of the White Bodies, and the Fat of the two Oriental Argent Vives, which Sulphurs are caught hold of, and retained by the Bodies.*

Salmon. All these are only Various Names, by which the Philosophers call the one thing, and under which they Cloud it. But the most Acute Ripley saith, it is Argent Vive, but not the Vulgar, that without which nothing that exists, is able to be. If therefore, there be nothing under the Sun, in which this Argent Vive is not, Our Hermes has not done absurdly, to call it by these Names; tho' possibly there may be some one thing, which may contain more of it, that which is more pure, also generous, and more ripe or perfectly digested, than all the other things besides. Authors say, it is chiefly found in the Roots of Metals, which Roots are in the Air, and the Tops of the Mountains. It behoves you therefore, to have a perfect and solid knowledge of this Argent Vive, before you attempt any thing in this Art. And this is to be Communicated only to the Faithful Disciples of this Science. Be diligent with your whole mind, consider, think, ruminate, volve and revolve, meditate and reason with yourself concerning it, and through the Divine Assistance, you will certainly attain to the knowledge thereof.

II.   Hermes. I say more, that this Sulphur does Tinge and Fix; and is contained and held by the Conjunctions of the Tinctures. Fats also Tinge, but withal they fly away, in the Body which is contained, which is a Conjunction of Fugitives only with Sulphurs and Aluminous Bodies, which also contain and hold the Fugitive Matter.

Salmon. He distinguisheth here between the true and d Philosophick Tincture, and the Fictitious or Sophisticate. The true is made of a Fixt and Incombustible Sulphur, for which Cause also, the Bodies are rendered fixt and Incombustible: for every Transmutation made subject to the nature of the thing Transmuting, and not of the thing to be Transmuted; it is needful therefore, that you make choice of the best Sulphur for this Work. The Vulgar is Foreign, for that it is deficient, Blackens, and Corrupts, having also a double superfluity, viz. an Inflamable substance, and an Earthy Faeculency. Therefore you must find out another, which is a simple Fire, and Living, and is able to Revivisie Dead Bodies, to bring them to the highest perfection, and to perfect them with the ultimate maturity. Such a Sulphur saith Avicenna is not to be found upon Earth, except in the Bodies of Sol and Luna. In Sol indeed is the highest of Perfection, because it is more digested and decocted; when therefore, the Tincture is prepared with this Sulphur, down below, in the Bottom of Obscurity, it is carried Gradatim up for the highest Glory, which the greatest splendor of spirituality, so that any Body whatsoever being melted with the Fire, it Tinges, and so firmly adheres to it, that it cannot for ever be any more separated therefrom. But the Sophisticate Tincture which is made from the middle Minerals, from burning Sulphur, Arsenical, Aluminous, and such like, are not able to defend either Bodies, upon which they are projected, nor yet their own proper substance from the violence of Fire, but together with the Bodies they flie away, and by the force of Fire Vanish into Air.

III. *Hermes. The order, method, management and*

disposition of the Matter sought after by the Philosophers, is but one, in Our Egg. Now this in the Hens Egg, is in no wise to be found. But left so much of the Divine Wisdom, as is seen in a Hens Egg, should be distinguished; we make in imitation thereof, a Compositum from the four Elements, joyntly fitted and compacted together.

Salmon. The Description of the Philosophick Egg is various, which the Philosophers divide into four parts, according to the number of the four Elements.
1. Putamen, the Shell which they make the Earth.
2. Albumen, the White, which is Water.
3. Pellicula, the Skin, which is Air.
4. Vitellus, the Yolk, which is Fire.

Some make only three parts thereof.
1. Vitellus, the Sulphur.
2. Albumen, the Mercury.
3. Putamen, the Salt.

Some again will have the Yolk to signifie Mars, Sol, and Venus; and the White, Saturn, Jupiter, Mercury, and Luna: and the Shell, the Firmament, and Earth, or Combustible Ashes; but to speak plainly, the. Shell represents the Philosophick Glass; wherein the Skin, the White, the Tread, and the Yolk, answer to the four Elements: Fire, Air, Water and Earth. Or rather the

Tread, Yolk, and White, to the three pure principles, Salt, Sulphur, and Mercury, or Spirit, Soul, and Body; that is, Fixity, Tincture, and Subsistence.

IV.   Hermes, Now in a Hens Egg, there is the greatest help that may be, for herein is a nearness of the Matters in their Natures; a spirituality, and gathering, and joyning together of the Elements, and the Earth which is Gold in its Nature.

Salmon. The Ovum Philosophorum, or our Mercury has in it self whatever is necessary thereto: We call it Our Mercury because it is reduced into one pure Homogene Body, where is

1. A Propinquity of Natures, as Earth, Water, Air, and Fire; or Salt, Sulphur, and Mercury.
2. A Spirituality, which is the formative faculty, the hidden Work-Master which brings the Stone to perfection.
3. A gathering together of the Elements, for that the Earth is made Water, and Air, and Fire, by Sublimation, and they are made Earth again; which Earth is Gold in its inward principle or Nature.

V.   Hermes. The Son saith to him, the Sulphurs which are convenient or fit for Our Work, are they Coelestial or Terrestial, Heavenly or Earthly? To whom Hermes answers: some of them are Heavenly, and some are from the Earth.

Salmon. This is a short Dialogue between the Father and the Son, Hermes makes answer to his Son, concerning the Sulphurs, that they are not of one or the same kind, but that some are of a Heavenly, and some are of an Earthly Nature, yet he confesses both to be Sulphurs: by the Heavenly is meant the Solar Sulphur; and by the Earthly, the Sulphur of Luna. For Sol is a Body Masculine, hot, fixt, red, and incombustible, which perfects Luna, who is Feminine, Cold, Volatile, White, and Combustible, exalting her to his own Glory and Splendor.

VI.   Hermes. The Son saith, Father, I think the Heart in the Superiors to prenote Heaven; in the Inferiors, the Earth. To whom Hermes saith: It is no so, the Masculine truly is the Heaven of the Feminine: and the Feminine the Earth of the Masculine.

Salmon. The Heaven is the Masculine of the Earth; and Earth is the Feminine of the Heaven: Heaven or Sol which is pure, fixt, and incombustible Sulphur, is the generating Seed; and Mercury which is the Magical Earth, is the Womb, or Feminine principle receiving the Seed, in which the Seed is kept, nourished, digested and brought to the Birth or Perfection: Even that in which it obtains, Spirit, Blood, and Flesh, viz. Fixity, Tincture, and Substance. The Earth or Mercury is the subject or receptacle of all the Celestial Radiations.

VII. Hermes. The Son saith, Father, which of these is more worthy, one than another, whether is the Heaven or the Earth? The Father answers: Both want the help of one another; but a Medium is proposed by precepts. But if thou shalt say, that Wisdom or the Wise Man does Rule or Command among all Mankind; to this Hermes: The indifferent or ordinary things are better with them, because every, Nature delights, or desires to be joyned in Society with its own kind. We find even in Wisdom itself, that equal things are joyned together.

Salmon. If by a Magical Matrimony, the Heavens and the Earth are Conjoyned, neither seems to be more worthy; for as the Earth cannot Generate without the Heavens, so neither can the Heavenly Influences multiply themselves without the Earth: But there is a mean

proposed, which he explicates in the following
Paragraph, which is the making the one equal with the
other, viz. by bringing forth, a new Off-spring out of
both, which shall excede Sol himself in perfection,
because it is more than perfect, and able to make the
imperfect Bodies or Metals perfect, which Sol himself
cannot do, and is also able to make the most imperfect
Bodies more than perfect, by multiplying their
Tincture a thousand times more than what they had
Originally by Nature, which is performed by Precepts,
saith Hermes, that is, by Art. And since every Nature,
Delights to be joyned with its own kind, you must be
sure to joyn Metalline Bodies only with Metalline
Principles, for Water joyned with Water cannot be
separated, no more can Silver being joyned with
Silver, or Gold with Gold, that is to say, their Seed.

*VIII. Hermes. The Son saith; But what is the mean
among them? To whom Hermes answers: To every thing in
Nature, there are three things from two. 1. The
Beginning. 2. The Middle. 3. The End, viz. Firstly,
the profitable and necessary Water. Secondly, the Fat
or Oyl. Thirdly, the Faces, or Earth, which remains
below.*

Salmon. By the Beginning, he means the Deundation of
the first Principles, for the Prima Materia must be

prepared and made fit for the Operation: the Middle, which are the Operations of the Work from the Beginning to the End: the End, which is the Perfection or consummation of the matter; these are the three things from the two Principles, Sulphur and Mercury: Or possibly by the three things from the two, he may mean the Spirit, Soul, and Body, i.e. Fixity, Tincture, and Substance, from the two Principles, Sulphur, and Mercury, being in Conjunction: Other Interpret it thus; by the two things he means Heaven and Earth, which cannot be Conjoyned without a Medium, (which is the Air) no otherwise than Soul and Body, which cannot be Conjoyned in one Body without the Spirit to Unite them. The Spirit then is the Legitimate Mediator of the true and perfect Conjunction, whether Natural or Supernatural. By the Heaven is understood the Soul: By the Earth the Body: By the Spirit the Uniting Principles; these indeed are the three things from the two, i.e. the two Principles, Sulphur and Mercury, the Spirit being Latent in them both. But however, our Hermes lest he should not be understood, has explicated them himself, viz. the profitable Water, the Fat Oyl, the Faeces or Earth. By the Water, is meant the Mercury; by the Fat or Oyl, the Sulphur, which by the Mediation of the Internal or Latent Spirit, are United into one Body, and make the Foeces or Earth.

IX. Hermes. But the Dragon dwells in, or Inhabits in all these things. And his Houses are the darkness and blackness in them; and by them he Ascends into the Air (from his rising) which is their Heaven: But while the Fume or Vapour remains in them, they are not perpetual, Permanent, remaining or fixt. Take but away the Fume or Vapor from the Water; and the blackness from the Fat or Sulphur, and Death from the Foeces: and by Dissolution, you shall possess a Tryumphant Gift, even that in and by which the Possessors Live.

Salmon. We have spoken now concerning the Heavens and the Earth, and their Matrimonial Conjunction, by a Medium, viz. the Air or Water, which we also call the Spirit; for the Water is nothing but the Air Coagulated; and the Air is the Vital Spirit, running through, and piercing all beings, giving Life and Consistency to everything, the very Agent which Ties the Particles of all Matter and Bodies together, and without which every Body, and Metal would fall to pieces, and become nothing but Dust and Ashes, even the smallest of Atoms: And this Spirit is that which moves and fills all things. It is the Philosophick Heaven, which in its prime Resolution or putrefaction, is wonderfully defiled, so that like the most Poisonous Dragon or Serpent, it destroys all things it touches; from whence it is said to have its House in Darkness and Blackness, and to possess Blackness, and

54

Clouds, and defilements, and Death itself: So long therefore as the Heaven shall be thus infected, it is impossible for it to return to its Pristine Nature, Simplicity, Purity, Fixity, and Permanency. By the Dragon then is signified this Black Matter Ascending into Air, which is difficulty done, by Reason of its thick glutinous and Ponderous Body, which would not tend upwards, unless it be first resolved by Force and Power of the Fire in a Philosophick Glass into an Aereal and Vaporous substance: being thus Dissolved by a frequent Ventilation of the Air or Spirit, it will be perfectly purged, and recover its Primaeval Nature of Heaven, which is the thing sought after.

CHAPTER VI

THE SEVERAL OPERATIONS BY, AND VARIOUS MATTERS OF, WHICH
THE STONE IS COMPOSED.

I. Hermes. Now the temperate Fat or Sulphur which is the Fire, is the Medium or Middle Nature, between the Foeces and the Water, and the through Searcher of the Water: The Fats are called Sulphurs, for between Fire, Oyl, and Sulphur, there is so little difference, that there is a propinquity, or nearness; because as the Fire does Burn, so also does the Sulphur.

Salmon. He here makes the Fire to be the Medium between the Sulphur and the Mercury, which Fire we have before declared to be both Internal and External: The first is Innate, in the Principles and Essential; the latter Elemental and Accidental; it is the through Searcher of the Water, that is the Stirer up of Internal Life and Efficacy; so that the Internal Fire may properly be called the Spirit of the Matter, which disposes the Particles of it to their Change: But the difference between this Fire, or Spirit, and Oyl or Sulphur, is so little, that we want fit Words to express it, but it is like the Spirit to the Soul, which are inseparable. But this is to be understood, that of Sulphurs, such only are to be chosen, which are the more near in their principles; the Sulphurs of Minerals are to be taken; not those of Vegetables or Animals: and of Minerals, that which is drawn from Mercury or Quick-Silver, Gold and Silver, which is to be Purified and exalted by some Power or Principle, which is without length, breadth, or thickness, viz. Incorporeal, and yet comprehends all those properties in it: without form or shape; yet comprehending under its formless being, the highest and most exact of Beauties; this is the Internal Fire of the Mineral Sulphur. And of these, the Sulphur of Mercury is yet said to be the most noble, because it is more at Liberty and free to Act, than the Sulphurs either of Sol or Luna, which are Fixt and bound up in a Dead or Life—less Body: The Sulphur of all the other Metals are yet more remote. And tho' they might serve the end, yet it is with more labour, trouble, and difficulty.

II. Hermes. All the Wisdom of the World is compre-hended within this, Learning the Art is placed in these wonful hidden Elements, which it does obtain, finish or compleat. It behoves him therefore, who would be introduced into this our hidden Wisdom, to quit himself from the Usurpation of Vice, to be Just and Good, of a profound Reason, and ready at Hand to help Mankind of a Serene and pleasing Countenance, Courteous in his Conversation to others; and to himself a Faithful Keeper of the Arcanums, being once revealed to him.

Salmon. The knowledge of this nearest Sulphur, and how to prepare and use it in this work, is the Summ of the hole Art; it begins, compleats, and finishes the whole thing. But how this Sulphur is educed out of a determined Matter, few Authors have yet taught. The Volatile must be first fixed, and the Wild Tamed, before you can Operate rightly, else you will never be able to hold the Matter, to Operate upon it; the Dragon must be then Mastered and overcome; being once Slain, you must endeavour to give it again a new Life, by raising it up into a new form, and restoring to it a new Volatility, to wit, the Life of Sol and Luna; which by a Conjunction, and Legitimate Digestion, with and in the Mercury of the Philosophers, gives to the new generation, a new Body, yet such a one as is Spiritual, Subtil, full of Life and Power, and able to penetrate into the most inward recesses of the most solid and compact Bodies, enriching even Vulgar Sol himself, with a thousand fold a greater Treasure than he contained before.

III. Hermes. And this know, that except you know how to Mortisie and induce Generation, to Vivifie the Spirit, to Cleanse, and introduce Light, how things fight and contend one with another, are made Colorless and freed from their defedations, or Spots and Foulness, like as from Blackness and Darkness, you know nothing, nor can you perform any thing.

Salmon. This Mortification, is intended of the first Life and form, without which you can do nothing, in order to Generation. You must make Alive by Killing, and Cleanse by first defiling or bringing to Putrefaction, and bring forth Light by first introducing Darkness. The two contrary Principles must first fight and contend one with another, and a Fatal War must be begun and carried on to the Destruction of the first form and Life, before the second form and Life can appear; and the Matter must be first made Colorless, that it may be able to receive the true Color, and Tincture. In Order to this, the Philosophical Calcination is the beginning of the Work, then Dissolution, that the sublimed Matter may receive its determination. To mortifie is to Dissolve any thing into the principles of which it is Compounded: Therefore saith Senior, there is no Generation without Corruption; and in this Putrefaction is the beginning of our Secret, which none but the initiated Sons of Doctrine and Philosophy do rightly understand. There must be a most close Conjunction or Matrimony, between the Superiors and Inferiors, between the Spirit and the Body, which is made by Ascention and Descention, through the Power of the Invisible Life.

IV.   *Hermes. But this you may know, that this great Arcanum is a Matter of so great Worth, that even Kings themselves shall Venerate it; the which Secrets, it behoves us to keep close, and to hide them from every profane and worthless Person.*

Salmon. That is, there is such a Vital Power, Strength, Efficacy and Virtue in this our Philosophick Tincture, that it is able to transmute all the Mercury in the World into fine Gold; and not only all simple Quick-Silver, but also the Mercury of all other Bodies as of Saturn, Jupiter, Mars, Venus, and Luna; which Power it exerts not only in those Inferiour Bodies, but is also able to transmute the whole Body of Gold into pure Tincture, and to exalt it to a thousand Degrees above what Nature has determined it to be.

V.   *Hermes. Understand also that our Stone is Conjoyned with, and Composed of many things, of Various Colours, and of Four Elements, which it behoves us to Divide and Cut in Pieces, and to Disjoynt them; and partly to Mortificie the Nature in the same, which is in it.*

Salmon. The Various things are Salt, Sulphur, and Mercury; the Body, Soul and Spirit; which Spirit is that which joyns the Soul and Body together. In Mercury it self there is a Salt, Sulphur and Spirit: The Salt of that Mercury is the Philosophick Earth, which is to be Dryed or Drained, Ploughed up, Manured and Cultivated; and the Sulphur thereof is the Internal Tincture, which Transmutes; but it is the Spirit or living Principle which gives the Permanency or Fixity, and without which all Bodies whatsoever would fall to Dust and Ashes; that is it which ties the Particles and Atoms of everything together. In Sulphur, there is a Volatile Body, but a Fixing Spirit, and a Rubine Soul which tinges: This Body in the Putrefaction is cast away, and only the Spirit and Soul, which are without Parts, or Proportion, without length, breadth, or thickness, without substance or corporeity are retained, and Conjoyned to the Mercury of the Philosophers by the Medium of Salt, in which Salt lies the depth of the whole Secret. This Salt is Vegetable, Mineral and Animal, from whence the Philosophers were wont to say, that their Stone was Threefold, according to that Ternary of Generations. In this Salt there is a Corporeity joyned with a Soul and a Spirit, that is with a Sulphur and Mercury spiritual, which are the Chains which tie all the Particles of the Body together. All these must be Cut in Pieces, separated and divided, in the beginning of the Work, which is done by joyning of the three together; this is a Mystery which only the true Philosophers and Sons of Doctrine can understand. You must joyn Body to Body, Soul to Soul and Spirit to Spirit, by which means you will make the separation; because the Soul will joyn with Soul, yet the Soul of the one will not joyn with the Body of the other, but separate.

VI. Hermes. And also to keep safe the Water and the Fire dwelling therein, which does contain its own Water, drawn from the Four Elements and their Waters; This is not Water in its form, but Fire, containing in a strong and pure Vessel, the Ascending Waters, lest the Spirits should flie away from the Bodies, for by this means are they made Tinging, and Permanent, or Fixed.

Salmon. That is the Mercury and Sulphur dwelling in the Salt; or the Spirit and the Soul dwelling in the Body, which is our Stone. The Fire (saith he) contains its own Water drawn from the Four Elements: That is, the Sulphur contains the Mercury drawn from its Original Fountains. This is not Water in Form but Fire. Nor is it Quick-Silver in form, but Sulphur; nor Spirit in form but Oyl, or Tincture, containing the

Clouds, and Ascending Waters, which are of a dry consistency or Body, sticking to the sides of the Glass, lest they should flie away in sublimation from the Bodies; by this means, being often iterated, at least three times, (but if it be six or seven times, it is yet better,) the Spirit enters into, and peirces and penetrates the Body, in Order to its Fixation: which at length is perfected which the highest Fixation and Tincture by the Fixing Oyl or Sulphur.

VII. Hermes. O Blessed Water in the form of Sea, which Element thou dissolvest. Now it behoves us, with this watery Soul, to possess a Sulphurous Form, and to mix or joyn the same with our Vinegar.

Salmon. Great is the Virtue and Power which dwells in the Aqua Philosophica, from whence it is called Blessed. For as common Water, washes away the Filth from things, and cleanses them outwardly; so this our Elementated Water, not only Dissolves Bodies, but also Washes away and Cleanses them inwardly from all manner of Defilemens and Impurities; and being joyned with the Philosophick Vinegar, brings forth from them their incombustible Sulphur, which by projection, tinges and transmutes all imperfect Metals into most pure finish Gold and Silver. This Water is the Key of the Art by which the Bodies are oftentimes to be opened, that is, they are to be Dissolved, and by the same to be again Coagulated, to be made more noble, pure, and perfect. So that no Foot-steps of Death, Blackness, Corruption, or Imperfection may any more remain in them. The preparation of this Water is known but to a few, nor do many attain to it, because the Well is Deep out of which it is drawn, nor do the Vulgar Chymists understand it. But whatever you do, you can do no great Matter without the help of Nature: and tho' Aqua Fortis and Aqua Regis and such—like, are usefull in their places, to dissolve and Tear Bodies into Atoms, yet are they Alien, and far from the true Aqua Philosophica, which has the Power to enter into the insides of Metals, whereas they, only divide them into many Superficies. And therefore say the Philosophers, the preparation of this Water is not to be Learned of Masters, but it must be taught by the Dictates of Nature her self.

VIII. Hermes. For when by the Power of the Water, the Compositum also is Dissolved, you have the Key of the Restauration, then Death and Blackness flie away, and Wisdom proceeds on to the Finishing of the Work.

Salmon. This Water does not Tear or Gnaw Bodies into Pieces and Bits, but it Radically Dissolves them, and reduces them into their Prima Materia, as they were in their Original Generation. Of this Nature are those Fountains & Springs in Hungaria, which have a Faculty of Transmuting what Iron soever is cast into them in to good Coper; and those other Fountains, into which if any Wood be cast, so as it remains but some certain time, by the Lapidescent Virtue of the Water; it is transmuted; through its whole substance into Stone; which Memorable and well known Powers and Operations of Nature in these particular things, are in part a demonstration, or at least an Argument to persuade one to the belief of other Operations and Transmutations in the Metalline Kingdom. Ignis & Azoth say the Philosophers are enough for the whole Work: Learn therefore from Nature, the preparation of this Azoth, or Water of the Philosophers; which Water being prepared, does with a simple Operation, through the help of Nature, gently boyling in a soft Fire, bring the work to a conclusion, and perfect the same. This Operation indeed, or simple Coction, is that which opens the Door into the Chambers of Life, making Putrefaction and Death, and blackness, and darkness to vanish and f lie away. This Water and this Fire, tho' simple, and simple in their Operation, yet are they hid; and known but to a few, for that they lead into the most recluse and abscondite recesses of Nature.

CHAPTER VII

THE OPERATIONS OF NATURE IN THE
AQUA PHILOSOPHICA AS IN
A SEED.

I. Hermes. Now know my Son, that the Philosophers chain up (the Matter) with a strong chain, or band, when they make it to contend with the fire: because the Spirits in the masked Bodies, desire to dwell therein, and to rejoyce there. In these habitations, they vivifie themselves, and dwell therein, and the Bodies hold, or contain them, nor from them can they ever be separated.

Salmon. The Bodies before than can be perfectly united with the Spirit, and joyned one to another in a strong Confederation, must first be purified and washed with Azoth and Ignis: for the washing is that which puts an end to the blackness; and the purification is made and continued till the White Elixir is made perfectly white, and till the red is made perfectly red; being thus cleansed and purified, the Spirit out of a natural pro-pension is drawn to the Bodies; in which being ardently inflamed, it immediately commixes with them, and they are conjoyned, with an indissoluble conjunction, under the Chains of which they remain inseparable for ever. Now this conjunction is not made by chance, but from the meer affinity which is between the Bodies and Spirit, for they both proceed from one fountain and principle, though of the two, the spirit, by reason it vivifies, and holds the Particles of the Bodies together, is much the more noble, the more excellent, and most powerful Agent.

II. Hermes. Then the dead Elements are revived, and the Compositum, or compounded Bodies are tinged and altered, and by wonderful operations, they are made permanent, or fixed, as the Philosopher saith.

Salmon. The Domicils of themselves remain dead, but the Inhabitants in them are alive. Now the Bodies of the Metals, are the Domicils of their Spirits; which when they are received by the Bodies, their terrestrial substance is by little and little made them, extended, and Purified, and by their Vivifying Power the Life and Fire, hitherto lying Dormant, is excited and stirred up. For the Life which dwells in the Metals, is laid as it were asleep, nor can it exert its Power, or shew it self, unless the Bodies be first Dissolved, Exalted, and turned into Spirit, (for that the Spirit does only Vivifie;) being brought to this Degree of purity and spirituality, and at length to perfection, by their abundant Virtue, they communicate their tinging property to the other imperfect Bodies, and Transmute them into a fixed and permanent Substance. This is the property of our Medicine, into which the Bodies are reduced; that at first, one part thereof will tinge ten parts of an imperfect body; then an hundred, after a thousand, then ten thousand, and so infinitely on. By which the Efficacy of the Creators Word is most apparently Evident, Crescite & Multiplicamini, encrease and multiply. And by how much the oftner the Medicine is dissolved, by so much the more it encreases in Virtue and Power, which otherwise without any more solutions, would remain in its simple or single State of perfection; Here is a Celestial and Divine Fountain set Open, which no Man is able to draw dry, nor can it be wholly exhausted, should the World endure to External Generations.

III. Hermes. O Beautiful and Permanent, or Fixed Water, the Formator of the Royal Elements, who having obtained (with the Brethren joyned with a moderate Government) the Tincture, hast found a place of rest.

Salmon. He does not call the Matter of the Stone simply Water, but a fixed Watery form, which whoso is ignorant of, knows nothing of the principles of this Science. This Fountain (saith Bernard Trevisan) is a wonderful Fountain of Virtue, above all other

69

Fountains in the whole World; it is as clear as
Silver, and of a Celestial Color. It is the Formator
of the Royal Elements; that is (as Bernard explicates
it) it draws to it self the King, who after 130 days,
it brings forth splendid, shining, and Crowned with a
Royal Diadem, who afterwards Adorns his Brethren, they
being first Purified in the same Fountain, and freed
from all their Internal Leprosie and Impurities:
By this he means, Concord and Peace is produced, and a
Stable Place of Rest, by which is prefigured, Tincture
and Fixation.

IV. *Hermes. Our Stone is a most pretious thing, yet
cast forth upon the Dunghil. It is most dear and
Valuable, yet Vile and the most Vile: (i.e. found
among the most Vile things.) Because it behoves us to
kill two Argent Vives together, and yet to Value,
Prize, and Esteem them viz. the Argent Vive of
Auripigment and the Oriental Argent Vive of Magnesia.*

Salmon. It is the most pretious, because it is the
Fountain of all Treasures, but cast forth upon a
Dunghil, because it is found in the Putrefaction of
the Matters, which is filthy and stinks like a
Dunghil; and so tho' it be most dear being perfected,
yet it is most Vile, as being found amongst the most
Vile things, in the midst of Corruption and

70

defilements. Our Stone is Composed of a double Argent Vive: The first of which Argent Vive's is Vile and Abject, and found in all places, in the Dunghil, in the High way, in Plains, in Mountains and in Vallies, and without which Man, is not able to live one Moment of an hour; for it enlivens all things, both Animals and Vegetables, even Herbs, Plants and Trees stand in need thereof; it preserves all things from Corruption, and every Mineral without exception. But would you know what it is; it is not Gold nor Silver, nor Gems, nor pretious Stones, these are things of great Price, and therefore not to be enumerated among those things, which are Vile and Mean. What is it then? It is Salt, but not the Vulgar Salt, with which Food is Drest, altho' that has one of the Qualities of this our Stone, viz. that of Dissolving; but it is Sal Petrae, Salt of the Rock, of the Rock I say, by which running Mercury, is transmuted into the best and most perfect of Metals, and the Flint into the most hard Adamant: but few will believe this, but such whom Experience and true Philosophy has taught, how it is found in all things, and by what Artifice it may be extracted out of them. This is that which without doubt our Author as under a Veil, calls Auripigment. And this is enough to be said concerning this Argent Vive, that it is Vile and most Vile. But the other which he calls Oriental Argent Vive of Magnesia, is most Valuable and Pretious; this is no spoke of Vulgar Argent Vive, nor of the Vulgar Magnesia. But by this Argent Vive, he means the humidity of the Mixture, which is the Radical Humidity of our Stone. By Magnesia, he understands the Total Mixion, or Compositum, from which this Humidity is extracted, and which Moisture is called our Argent Vive; which Humidity does indeed run in the Fire and in the same does dissolve the whole Compositum, and also congeal it, makes it grow Black, makes it grow White, and also makes it grow Red, and in the end compleatly perfects it; and it is that which does all in all, being a most pretious Treasure to such as know it, and possess it. This Magnesia is the Power and Virtue of our Stone, which like an Universal Magnet draws all things to their Center, whether in the superior or inferior Worlds. And the greater part of this Secret Arcanum lies rather in this pretious Magnesia, than in the former Vile Argent Vive: of which, tho' the Philosophers have variously hid it under Clouds and Veils, we have here said enough.

V. Hermes. 0 Nature, the greatest, the Creator of Natures, which makest, containest, and separatest Natures in a middle principles Our Stone comes with Light, and with Light it is Generated, and then it Generates or brings forth the Black Clouds or Darkness which is the Mother of all things.

Salmon. Universal Nature is but one thing, which is

the very principle of Motion and Rest, and which, as Hermes saith, is the Creator of Nature, or the producer of all things. But God Almighty is the Supream Work Master, and great Architect of the whole World, who created and brought forth this Universal Nature, that according to his Determination it might bring forth all other things in a middle principle, which is that of Generation, by a proper and specifick Power. So if Grain be cast into the Ground, God Almighty by his Instrument, Nature, causes it to be Spring, and Grow; but this Education is in the middle principle, viz. the inward and latent Life, and energetick Spirit which specificates it to its own right & particular form, bringing forth from Wheat, Wheat and not Barly, Rye, Oats or Pease, so if the Seed of Gold which is Light, be Sown in a proper and sit Earth, Meliorated and made fit for the purpose, Nature by Virtue of the energetick Spirit or Light, lodged in the Seed of Gold, specificates that production, and makes it bring forth Gold again, 10, 100, or a 1000. fold, according to the goodness of the Earth in which it is Sown: But before it is brought to perfection, the Light must be Eclipsed, the Seed must Die, Corruption must prevail, and Darkness must Over-spread the Face of the Earth: By which say some Philosophers, nothing else is understood but the Magnesia of Saturn, now Saturn by the Greeks is called Chronos, that is Time, in which all things are produced, and the Magnesia it self, which is the

Mother and the Generatrix of our whole Work.

VI. Hermes. But when we Conjoyn the Crowned King to our Red Daughter, and in a Gentle Fire, not yet too great or hurtful, she does Conceive, and will bring forth a faithful and excellent Son; which does feed with a little Heat, and nourishes the permanent or fixed Matter, making it to abide even the greatest Fire.

Salmon. The Queen or the Red Daughter of the Philosophers is Luna, to wit, the Metalick or Philosophick Luna, which now puts on the Masculine Nature, by being Conjoyned to our Sol, our Crowned King, and she brings forth a Son, which the Philosophers call their Mercury. This is wonderful, that the Parents who before were the Nurses and Feeders, are now by the same Nursed and Fed, but it is so in this Our Work. It is nourished with a gentle Heat, (not in the Vulgar way of Decocting) answerable to that of the Heavenly Fire or Sun. And it is fed 10 or 12 times with its proper Food and Drink, which is the Mercurial Bread and Water, by which it grows, increases, and is brought to perfection, resplendent in Glory like the most sparkling Fire. This Son ought indeed to be fed to Satiety, even so many times till it neither Hungers nor Thirsts any more, then it is

Tinged and Fixed for ever.

VII. Hermes. But when you send forth the Fire upon the Leaves or enfoldings of the Sulphur, the Boundary of Hearts does enter in above it, is washed in the same, and his Putrefied and stinking Matter is extracted; then he is altered or changed, and his Tincture by the help of the Fire remains Red as Flesh.

Salmon. Every thing which lives, lives by Virtue of its inward Fire or Heat; and Sulphur contains within it a hidden Fire, which by the External Fire is excited and stirred up; Life made manifest begins to live, and that which before was hid in the Sulphur, now exists and is made manifest; it is the business of the Fire, not only to Vivifie, but also to Depurate, and Segregate the things which are Heterogene, till (being separated) there appears at length in the Foeces a most pure and Rubicund Tincture of the Color of Flesh newly Killed and Bloody: This is the Blood of the Green Lyon, which the Philosophers speak of; and it is said to be Green, not for any external Green Color, but from its Viridity or strength of Life. The Tincture: is like Bloody Flesh new Killed, or Blood yet flowing and moist, which then is said to have attained the Degree of perfection. And as Flesh is nothing but Blood Coagulated; abounding with a ful

75

vigorous and perfect Spirit; so also Our Tincture is
nothing but the Coagulated Blood (which Blood is the
boundary or satisfaction of Hearts) even the Blood of
the Green or Virescent Lyon, vastly repleat with a
Fountain of Vital Spirits.

*VIII. Hermes. But Our Son the begotten King, doth take
his Tincture from the Fire; And Death, and the Sea,
and Darkness fly away from him.*

Salmon. Now if you know what this Tincture is, Our
Hermes here tells you very plainly, that it is made
and remains Red by the help of the Fire; and again,
Our Son the begotten King Sumit Tincturam exigne, doth
take his Tincture from the fire, from whence plainly
it is taken; the fire is that strong Fortitude, or
invincible strength, which brings forth this Tincture,
or true Viridity of the Lyon. Whatsoever tincture
flies away from the Fire is Immature and Imperfect;
nothing can be right and perfect but what can endure
the strongest Tryal of that Element; and therefore by
consequence the Tincture is to be so long nourished by
the Fire, till it comes to the height of Perfection.
And thus our tone, which before, viz. in its
beginning, lay in Death, and was drowned in the Sea or
Waters, and surrounded with Darkness, which was the
Corruption of the Matter, is by the Power of the Fire,

with a gentle Coction, assimulated to the Nature of the Fire, and at length wholly turned into Fire, where it dwells as in its proper Matrix or Element, and in the same only rejoyces and is delighted, till by length of time it is coverted into a Quintessence the true Philosophick Tincture, and so Triumphs over Death, the Sea, and Darkness as having really Conquered them, becoming a Medicine for the Bodies both of Metals and Humane kind.

IX. Hermes. The Dragon flies from the Beams of the Sun, who observes the Holes or passages, where they enter in; and our Dead Son Lives. The King comes from the Fire, and rejoyces with his Wife, laying Open the hidden things; and Obscured Virgins Milk.

Salmon. The Dragon here signifies the Earth, which is Black, blacker than Black: Now Serpents and Dragons delight rather in places under Ground, Holes of Rocks, and obscure Dens, than abroad in the Open Air and Light of the Sun, and therefore they avoid the shining Sun, viz. the Spirit of the Fire. This Dragon must be inclosed in a Vessel, little and round, well defended and Luted, and close stopped up, and then exposed to the Heat of the Sun for one Philosophick Month, or space of 40 days, in which time it is destroyed, the parts being wholly Dissolved, by the deadly stink of which the King's Son happens also to be Killed: Both being therefore slain and put into another clean Glass, are put in the Heat of the Sun for other 40 days, or other the like Heat, and in that space of time, the Dead Son by little and little is revived and restored to life; but the Dragon by the same heat, being vehemently over-born is wholly Consumed and Reduced to nothing. Now the King seeing the Heat of the Sun to be too weak to Work a Total deliverance from the Poyson of the Dragon, prepared a Bath for him self and His Son to wash in; in the mean time the Virgins Milk is brought to its Whiteness, with which the Son is Copiously Fed, and the hidden Tincture is brought to light, and advanced to the height of its Glory.

X. Hermes. Now the Son Vivified, or made to Live, is made a Warrior of the Fire, and superexcellent in his Tinctures; for the Son has got the Blessing, having also the Root of the matter in him.

Salmon. The Father can never Desert the Son, for the Son is of him and from him, participating of his Life and substance: and is like unto him in all things; and by this new Generation is made patient and strong, able to endure the most vehement and lasting fire, without the least Diminution or hurt, to its substance. The Son has got the Blessing, that is, the Tincture and Fixity of parts. And the Root of the Matter is in him, that is, the Prima Materia, the Aurisick Seed, out of which the Golden Tree of the Philosophers is said to Spring and Grow, bringing forth much Fruit.

XI. Hermes. Come ye Sons of Wisdom, and rejoyce: be ye glad and exceeding joyful together; for Death has received its Consummation, and the Son does Reign, he is invested with his Red Garment, and the Scarlet Color is put on.

Salmon. Having Conquered and overcome this horrible Monster; this terrible Dragon, this Poysonous Serpent, this Malign and prof ligating Spirit, this Putrefaction, Corruption, and Darkness, this almost invincible Death, and brought forth, a new Offspring to Life, Glory, and Perpetuity; full of Spirit and Power, of Sulphur and Tincture, even the highest Redness, enjoying a most fixt substance, always encreasing ad infinitum, which is the Reign and Dominion of this new Birth, Clothed with his Red Garment of Scarlet Color; I say, having done all this, Hermes advises us to rejoyce and be glad, yea exceeding joyful; for this is the final end of Care, and Trouble and Sorrow; making Rich with a Treasure that can never be Consumed, Wasted, or brought to an end.

CHAPTER VIII

THE PHILOSOPHICK RIDDLE LAID DOWN AFTER A NEW MANNER.

I. Hermes. Now understand that this Stone cries out, nourish and perfect me, and I will reward you; give to me mine, that which is my own, and I will bountifully recompence you.

Salmon. It ought to be nourished and brought to perfection with proper Food of its own Nature, that is with Fire and Azoth; with Spirit, and the Virgins Milk: The true Aqua Philosophica, which gives it substance, Tincture and Fixity. This is its own, and is like Leven made out of the same Paste, which for ever afterwards levens the whole Lump: if you do this, you will reap the Recompence of reward, the Fruits of your Labour, Riches, and Honour, and Glory, and every good thing.

II. Hermes. My Sol, and my Beams are most inward, and secretly in me: my own Luna also, is my Light, exceeding every Light; and my good things are better than all other good things.

Salmon. That is, our Mercury contains inwardly within its own Bowels the Aurisick Seed, but it is most inward, even Centeral, so that it seems to be hid from the Vulgar Eye; this substance must be turned the inside outward, which can be done no ways but by Putrefaction, that the solar Sulphur may be made to appear. It contains also in itself Luna, which is unripe, untinged Gold; yet it is said to be the Light, as being the Seed from which the Philosophick Tree, the good things of our Stone do proceed; the Aurora, the Morning of the Glorious Day.

III. Hermes. I give largly and plentifully: I reward the understanding with Joy and Gladness, with Delight, with Riches and Honour and Glory: And they that seek after me, I give them opportunities to Know and Understand, and to possess Divine things.

Salmon. This is a Prosopopoeia, shewing the liberality of the Donor, the Infinite and immense Treasures (as it may in a sense be said), with which all the worthy Searchers after this wonderful Mystery are Blessed: the Rewards are Honour, and Glory, and Treasure, infinitly exceeding that of Kings: The possessors of which flight, and undervalue all temporal things, in comparison thereof, and despise their uncertain, and fading satisfactions for that this can never be lost, nor spent, never be Exhausted, or consumed, but remains as a fountain always running, an Eternal Spring for ever.

IV. Hermes. Now know that which the Philosophers have hidden and obscured is written with seven Letters. Alpha and Ida, follow the two: and Sol in like manner follows the Book (of Nature) notwithstanding, if you are willing, or desire that he should have the Dominion you must watch the motions of Art, to joyn the Son to the Daughter of the Water, which is Jupiter: This is a hidden secret.

Salmon. Under the notion of seven Letters (signifying the seven Planets, and the seven Metals answering to them) Hermes has hid the Secret; and in the following words, as in an Aenigma, has lockt up the Great Mystery, so that it is hard for any one positively to give a true interpretation. Tho' by Alpha and Yda, tis probable he means the Spirit and the Soul, which follows the two, Mercury and Sulphur, which is the Conjunction of Sol and Luna, a Mercury and a Sulphur fixed and Incombustible. And by Sol following the Book (of Nature) is meant the natural Course of Generation, which is the same thing in Metals as in all other things. By the Daughter of the Water or Jupiter is meant Mercury, and by the Son, a Sulphur fixt and incombustible, such as are Sol, and Luna: in the Conjunction of which Mercury and Sulphur lies the whole Secret. These two when conjoyned are but one thing: but there are two several kinds of fixed Sulphurs; the one and the more excellent is Solar and Red; the other Inferior to it, is Lunar, and White, out of which are made the different Ferments for the Elixirs White and Red.

V. Hermes. Auditor understand, let us then use our Reason: Consider what I have written with the most accurate Investigation, and in the Contemplative part have demonstrated to you. The whole manner I know to be but only one thing.

Salmon. He which easily believes, may easily be deceived, and therefore he advises us to use our Reason; not to take things according to the Naked found of the Letter, but to consider the weight of the Matter, the Power of the words, and the attendant circumstances to the same; what he has here most subtly investigated, you ought to consider with a profound contemplation: However, the Root of the matter he positively and plainly tells you is but one thing, which is the Aqua Philosophica.

VI. Hermes. But who is it that understands the sincere investigation, and inquires into the Reason of this Matter? It is not made from Man, nor from any thing like, or akin to him, nor from the Ox or Bullock. If any living Creature can joyn with one of another Species, the thing is Neutral indeed which is brought forth.

Salmon. Our Hermes has given us the sincere investigation of Matter, the true and right reason of the Operation, consentaneous to the Laws of Nature, but in some things he has left us in the Dark, at least to the use and exercise of our Reason and Judgment. And tho' he does not expressly say what the matter is taken from, yet he plainly tells you what it is not taken from; you cannot gather Grapes of Thorns, nor Figs of Thistles, is a Dictate from the Oracle of Truth; and so Hermes tells you, a Metalline Body and Substance cannot be taken from an animal being: But Man brings forth Man, and Beast, Beasts; the Ordination of God in the Creation of things remains inviolable; and if different Species of the same Genus mix together, a contamination of both the Species follows; this is plain to the senses: the same thing happens also in Metals.

VII. Hermes. Now Venus saith, I beget the Light, nor

85

is the Darkness of my Nature; and unless my Metal be dryed all Bodies would cleave unto me; because I should make them Liquids: Also I Blot out, or Wipe away their Rust and Filthiness, and I extract their substance: Nothing therefore is better than me and my Brother, being Conjoyned.

Salmon. This is spoken Allegorically, because Venus as the Morning Star is the Harbinger or Forerunner of the Sun Rising. Where is Light there is Life, the Light being the Vehicle of the Life: There is nothing in Rerum Natura, which is not brought forth by the help of this Light, viz. by a Natural Generation: Metals are thus produced in their Mines: But this Light is not found in Metallick Bodies, because of their too great dryness and Terrestreity; and therefore because of the moistness of Venus, they would gladly stick to her. This moist Metal Venus, which is neither Copper nor Brass, is endued with Lucidity and Splendor, and with a Fiery Virtue and Power, by which is melts Bodies, as if it was a Fire of Coals; but it melts or liquifies them not simply, but by melting them, washes away their Rust and Corruptible matter; that is, extracts and brings forth to light their Purity and incorruptible substance, even their inward and hidden Tincture. What is done then? Truly if Venus and her Brother copulate together, and at length, by the Craft of Vulcan, are taken and held bound together (by some invisible Power or Spirit) in Chains; she will be impregnated, and after a Revolution of ten Months, bring forth a Son more Noble and Excellent than the Parents. This is the pretious Stone, a Pearl of great price, the invaluable Treasure, which even the Kings and Princes of the Earth, and the Great Ones of the World seek after; but it is hid from their Eyes, being only the proper Inheritance of the abject and humble in Spirit, who are the true Sons of Wisdom.

CHAPTER IX

THE LAST ACT, OR CONCLUSION OF THE THEORY OF THE PHILOSOPHICK TINCTURE.

I. Hermes. But the King, and Lord, or Dominator, to the Witnesses his Brethren saith, I am Crowned, and Adorned with a Royal Diadem, I am cloathed with the Royal Garment, and I bring joy and gladness of Heart.

Salmon. By the King is meant Gold; and by his Brethren, the other inferiour Metals, which all possess the Kingdom in common, the supream power of which resides in Sol alone, for that he sustains himself in the fire without hurt, even to the longest period of time. By the Royal Diadem he means Fixity; and by the Royal Garment Tincture, even the red Tincture of the Stone, which as Ferment or Leven, Leavens all the inferior Metals, and transmutes them into its own Nature and Property, and this by the help of our Mercury.

II. Hermes. And being overcome by Force, I made my sub-stance to lay hold of, and to rest within the Arms and Brest (i.e. the Body or Womb) of my Mother, and to lay hold and fasten upon her Substance: making that which is Visible to be Invisible, and the hidden Matter to appear: for everything which the Philosophers have Vailed or Obscured, is Generated by Us.

Salmon. That which is thus overcome by Force is Sol; that is, it is dissolved and its Body Opened, and made to joyn and Unite with Mercury, which is the Womb in which the solar Seed is Sown, which is the Mother thereof: in which Womb being digested and Ripened, it lays hold of the substance of Mercury, fastens upon it, and converts it into its own Nature: Thus Sol which before was Visible, its substance being attenuated, is made invisible, and a Spirit; and that which was before hidden and invisible, is made to appear, which is the Internal Soul and Spirit; that is, Tincture and Fixity, which by Virtue of the Ferment is put upon Mercury, whereby the Vailed or Obscured Matter is Generated, which is the substance of our Stone, whereby a Door is Opened into the Chambers of infinite Treasures.

III. Hermes. Understand these words, keep them, Meditate upon them, and enquire after nothing else: Man in the beginning is Generated of Nature, whole Bowels or inwards are Fleshy; and not from any thing else. Upon these words Meditate; and reject what is superfluous (to the Work).

Salmon. With what Vehemency and Earnestness does Hermes here speak, as tho' the whole Mystery lay in these words: And truly not in Vain does he bid understand them, keep them meditate upon them, and to enquire after nothing else: You cannot gather Grapes of Thorns, nor Figs of Thistles: As a Man Begets or Generates a Man, and a Beast a Beast, and as every Hearb, and Plant, and Tree are produced from their proper Seed; so in the Metalline Kingdom, Metals are only produced from Metalline Seeds or Roots, cast into a proper Womb, which is the Philosophick Mercury, the Earth whence they draw their Nourishment, and by which they Grow, Encrease, and proceed on to Perfection. All other things whatsoever are Vain and Fruitless.

IV. Hermes. From thence saith the Philosopher Botri is made; from the Yellow or Citrine, which is extracted out of the Red Root, and from nothing else; which if it shall be Citrine, thou hast sought it at the Mouth of Wisdom, it was not obtained by thy Care or Industry: You need not study to exalt or change it from the Redness: See I have not Limited you, or Circumscribed you under Darkness; I have made almost all things plain to you.

Salmon. By Botri he means the two Stones, the White; and the Yellow or Red, which are extracted out of the

White and the Red Roots, viz. out of the Sulphur of
Nature. That which Whitens, the same also makes Red;
and the same that Kills, the same also makes Alive:
(Qui mecum moritur, mecum oritur.) But this is true,
only of the great Work it self; and not of any Branch
thereof; in particular Works and Operation, you must
have particular Ferments, which must be taken from
Luna for the White, and from Sol for the Red, as the
Arabian Geber has at large and plainly taught us.
Nature does only and alone conjoyn and separate, and
all its Operations are subtil and spiritual; but if
you will be Wise above Nature, you shall certainly Err
and suffer an irreparable loss: And having once
brought it to the fixed Redness; there is nothing
beyond that, in that is the Ultimate perfection, where
you must take up your rest.

*V. Hermes. Burn the Body of Laton or Brass with a very
great Fire, and it will give you Gratis what you
desire; it will Stain, Dye, and Ting, as much as you
can with it, and that with Glory and Excellency. And
see that you make that which is Fugitive and Volatile,
or flying away, that it may not fly, by the means of
that which flies not.*

Salmon. By the Body of Laton or Brass, and by that
which is Fugitive, he means the Philosophick and

Volatile Mercury, which by a Sulphur fixed and incombustible (such as is taken from Luna and Sol) is to be fixed in the Fire, so as it may rest and remain therein tho' most Vehement, and Fusory, or in the strongest reverberation, without the least Diminution, Detriment, or Corruption. But the Mercury is fixed by the Spirit of the Sulphur, not by its Corporeity; the Corporeal Particles only give form, and convey the Spirit to the Mercury, which could not be brought to it, in or by any other Vehicle. By the Fire, all the Heterogene or impure parts of the Mercury are destroyed, the pure left behind, and held fast by the Power of the fixing Spirit, which otherwise without the assistance and help of that Spirit would have vanished also: This Volatile substance it seizes upon, changes or transmutes, and fixes, that is, brings over into its own property. This tho' a Spirit contains in it the highest fixity, and its Body Being Opened, is the Sulphur, or Seed which must be sown in the Philoso-phick Earth or Mercury, (as we have often said) that it may there Die, and resume a new Body, a thousand fold more in quantity than its own Power of the Energetick Spirit will be made to live, spring up and grow, to be a Tree, of the first Magnitude, bearing Golden Apples, whose Seed will be and remain in itself for ever, and bring forth a new to infinity of Generations: Its perpetual new Substance or Body being made out of the Substance of the Mercurial Earth, by the Power of the transmuting or Vegetating Spirit and

Soul.

VI. Hermes. And that which rests or remains upon a strong Fire (is fixed,) and is also a strong Fire itself: and that which in the heat of a strong or boyling Fire is corrupted, or destroyed, or made to fly, is Cambar.

Salmon. By Cambar also he means the self same thing, to wit, our Volatile Mercury, in its Corruptible State: or rather the Corruptible and impure part thereof, which must be corrupted, or destroyed, and made to fly away, that which is pure and will not fly, may appear and remain; but the purifying Fire must be known, in which the great Secret of the Operation lies; and without which nothing can be done, which Fire, as we have formerly said, is two fold, viz. Internal and External, the latter being used only to excite the former.

VII. Hermes. And know ye that Our Aes, Brass or Laton, is Gold, which is the Art of the permanent or fixed Water; and the Coloration of its Tincture and Blackness is then turned or changed into Redness.

Salmon. That is, Our Gold or Stone, or Tincture is the product of the permanent or fixed Water, by which he means the Philosophick Mercury impregnated with the Spirit of the fixed and incombustible Sulphur. And by this you may perceive he puts a difference between the Aes, Brass or Laton, which is made by this permanent Water, and the Corpus aris, or Body of common Brass. Now the Aqua permanens is that which contains in itself the Tinctures of all Colors, Black only excepted, which is taken away from it, for that it is a sign of imperfection and impurity: By this Water alone Mercury is turned or changed into the true Red, that is, into the Tincture of Sol. But to take away its Corruption, and to reduce it into the incorruptible and fixt Nature of Sol, that must be done by Sol alone, and not by any corruptible and Foreign Matter or Substance, for that Sol contains in himself the Seeds of fixity and Tincture, which no other Body in the World does besides. But to make Sol do, or perform these things, its Body must be opened, prepared, and made fit for this purpose, by Virtue of the Aqua permanens, or Aqua Philosophica.

VIII. Hermes. I confess that through the help of God, I have spoken nothing but the Truth: That which is destroyed must be restored and renewed, and from thence Corruption is seen in the Matter to be Renovated, and from thence the Renovation appears: And on both or either side, it is the sign of Art.

Salmon. He has hitherto been teaching you the first
part of the Work, which is the Destruction of the
first Birth and Life; concerning which he assures us,
he has spoken nothing but the truth: Our Mercury must
be undone, and unmade, that is, corrupted and
destroyed, and brought through Putrefaction into a
pure and Limpid Water, that it may be able to peirce
the Metalline Bodies: from which State, by Conjunction
with a pure, fixt, and incombustible Sulphur, and by
Virtue of a subtle, living and fixing Spirit,
invisible, without length, breadth, or thickness,
(which Spirit is the Philosophick Fire,) it is to be
renewed and regenerated; the Water is to be dryed up,
the spiritual is to be made corporeal; the thin to be
made thick, the Volatile to be made fixt; and the
changeable Colors reduced to a Unity and Permanency,
either White or Red, according to the Order and Root
of the Operation; one and the same Mercury does
corrupt and destroy the Bodies, and again exalt,
perfect, and fix them; The Matter of Our Stone is but
one, and therefore nothing can be more Alien from the
Art, than to fetch it from many things; Nature is not
mended or made better, but by a Nature of its own
kind; as Vinegar makes Vinegar; so Our Art begins with
Mercury, and with the same Mercury it is Finished. It
is a kind of Proteus, which, creeping upon the Earth,
assumes the Nature of a Serpent; but being Immersed in

Water, it represents a Fish; presently taking to it self Wings, it ascends a loft, and flies like a Bird; yet notwithstanding it is but one and the same Mercury; with this the Artist does Work, and with it he transacts all the necessary Operations of our Stone, being fit and proper for them all, viz. for Putrefying, Distilling, Coagulating, Mortifying, Vivifying, Subliming, and Tinging, without which seven Operations you labour wholly in vain. Till you have putrefied the Matter, you have not made one step in the true way; but that being done, you have accomplished the first sign of the Art, as Hermes testifies.

CHAPTER X

THE PRACTICAL PART OF THE PHILOSOPHICK WORK.

I. Hermes. My Son, that which is born of the Crow is the beginning of this Art.

Salmon. The Crow is the Blackness and Darkness of the Matter being Corrupted: Now nothing was ever Generated or brought to light, which had not its beginning from blackness and darkness, ex nocte Orphei, i.e. from principles Invisible; for so it is said concerning the Creation of the great World. In the beginning when God Created the Heavens and the Earth, the Earth was empty and void, and darkness was upon the Face of the Deep; and God said, let there be Light and it was so: From whence we may gather, that Darkness was Prior to Light: and so it is in this our Philosophick Work, and altho' it is commonly thought that the darkness or Dark principle is taken for the true Seed of things, yet it is no such thing, but only certain Rudiments, or rather the Domicil wherein the true Seeds of things dwell: Nor is the Spirit by it self the Seed of things, nor yet the Corporeal Particles by themselves; but a certain portion of Spirit joyned with a fit proportion of Idoneous Matter conjoyned with an Eternal Soul; which in the beginning of our Work is to be Putrefied, and made blackness and darkness, that the whole Corporeal form may be made spiritual; and the Seed which before was Corporeal and Visible, or a Spirit joyned with a Soul and a Body, may become wholly spiritual: From this third, Immixt, Incorporeal, and Invisible Seed, as from the Crow, in the blackness and darkness of the Night, is our Stone, the true Seed brought forth, which, saith Our Hermes, is the beginning of this Art.

II. Hermes. See here, how I have obscured the matter spoken of to you, by a kind of Circumlocution; and I have deprived you of seeing the light (by giving you too much light:) And

1. This dissolved,

2. This joyned,

3. This nearest and longest, I have named to you.

Salmon. He tells us he has not nakedly demonstrated the whole thing to us, but he has Indigitated the Matter with what sincerity he could, Circumscribendo, by a certain going about or Circumlocution, which the Sons of Art by thinking and Meditating upon, may at length happily find out. The Philosophers say, there are three several Birds, which from the Name of Hermes, they call Ayes Hermeticoe which fly by Night without Wings. The first is Coruns the Crow or Raven, which from its blackness is said to be the beginning of the Art; and is of the Nature of the Element of the

Earth. Another is the Swan, and is so called from the Whiteness in the middle, and answers to the Element of the Water, in which the Swan is Conversant. The third is the Eagle, which is the Oleum sen Sulphur Philosophorum, and answers to the Air, for that it flys longest in the Air, and nearest to the Sun. But that you may not be deceived by Names, these Birds, answer to so many Spirits, or rather to one only Spirit under that threefold appearance, or manifestation.

III. Hermes. Roast them therefore, then boyl them in that which proceeds from the Horse Belly, for 7, 14, or 21 days; that it may eat its own Wings, and kill or destroy it self. This done, let it be put in Petta Panni, and in the fire of a Fornace, which diligently lute and take care of, that none of the Spirit may go forth: And observe, that the times of the Earth are in the Water; which let be as long as you put the same upon it.

Salmon. Hitherto he has for the most part, delivered the Art Theorically, now he comes to the Practical part, ordering the matter (before demonstrated in divers manners) to be roasted, and to boil it in Horsedung, for a certain number of days. There is a time of digestion, which is the prime, or first

Affation, or Decoction, with a fire weak and soft, like that of horsedung, which is sufficient for the first degree of Digestion: This being done, the Dragon will eat his own Wings, and kill or destroy himself, that is, the matter will begin in the Terra Philosophica, to be dissolved and corrupted. Then after the time of the solution is absolved or compleated, the heat of the fire is by little and little to be augmented, and the matter to be decocted in a Philosophick Fornace or Athanor, with a continual fire. But the Vessel which must hold the matter, must be exactly sealed, left the Mineral Spirits, (which have a most vehement and penetrating fire) should transpire, or go forth, and leave the dead Body: This may be done with Lutum Sapientiae, which you may prepare after this manner. Take Glue dried into powder, one ounce, Barly flower two ounces, green Wood Ashes, Salt, Calx Vive, Sand, Crocus Martis, or Caput mort. of Vitriol, ana one ounce, all being in fine powder, let them be mixed with juice of Comfrey, and Whites of Eggs, to the just consistency of Lute: with this the Mouths and junctures of the Vessels must be stopt and closed, so that the least Spirit or Vapour may not go forth.

IV. Hermes. The matter then being melted or dissolved and burnt, take the brain thereof, and grind it in most sharp Vinegar, or Childrens Urine, till it be obscured or hid: this done; it does live in putrefaction.

Salmon. Our Stone contains secretly or hiddenly in its self all the Colors of the World, which are not manifested, unless it be first melted or dissolved. As often therefore as it is melted in the fire, so often a new color arises from it, till all the colours are vanished, and the whole matter is reduced to ashes: And in these Philosophick Ashes is the Phoenix hidden, and out of them will it arise with glory and splendour; at first weak like a Worm, which in success of time will become a Bird, even the most glorious Phoenix. By the Brain thereof, he means the Spirit: But here he calls the Ashes the Brain, Metaphorically; for as the Brain is the Seat of the most pure and subtil Animal Spirit, in an Animal; so these Ashes are the place of the most subtil Mineral, or Metallick Spirit, and the matter in which the said Spirit is hidden, even the most noble, and most pretious Spirit of this whole greater World. By the most sharp Vinegar, or Childrens Urine, he means the Acidity, or Spiritual juices of the Metals, or Metallick bodies: And by grinding the Ashes therewith, he means dissolving them therein, which is the Philosophick way of speaking: And this dissolution must be so long, till it putrefies, and the first color of the operation appears, which is blackness; which color must twice appear. The Stone must become Black twice, twice White, and twice Red; the cause of which is but one only, for that the putrefaction is twice repeated; and therefore it is said the second time to live in Putrefaction; that is, being once corrupted and putrefied; the second time it does putrefie. By the Brain (as I have said) is understood the Spirit, or the most subtil Mineral substance dissolved in the Radicated Vinegar of the Philosophers; if you know not the preparation or rectification of this Radicated Vinegar, you know nothing of the true Philosophick Menstruum, or dissolvent; there is no other Aqua Vitae Metalica, Aqua Vitae Mercurialis, Aqua Lapidis, but this Acetum Radicatum, for that it contains in itself all things necessary for this Work.

V. Hermes. The Dark Clouds will be in it, before it is Kill'd; let them be converted into its own Body; and this to be reiterated as I have described: Again let it be Killed as aforesaid, and then it does Live.

Salmon. That is to say, while the Matter is in Dissolution and Putrefaction, in Killing but not Killed, the Clouds like a Tempest, will arise, which is an effervescence caused from the contest of the contending principles, as is evident in all forts of Fermentations. These Clouds must revert again, and be converted into their own Body; and this Work must be so often reiterated, till no more Clouds arise, viz. till the Dragon is wholly Slain. This done he must be restored to Life again, and made to live, and then killed again, as aforesaid, and then it does live, (as we have demonstrated in the Explication of the former Paragraph,) even in Putrefaction, from which it must at length (by the order and course of the Operations) be freed and brought to its Ultimate Perfection.

VI. Hermes. In the Life and Death thereof the Spirits Work: For as it is Killed by taking away of the Spirit; so that being restored, it is again made Alive, and rejoyces therein.

Salmon. The Spirit is used both in Killing of it, and in the making of it Alive again: but this is by some doubtfully understood, whether it be meant of the innate or indwelling Spirit only, or of that Spirit joyned with another Metallick Spirit, because he uses Spirits in the Plural number: However this is certain, that as Death is induced by taking away the Spirit; so Life is retrieved, by restoring it again.

VII. Hermes. But coming to this, that which ye seek by affirmation, ye shall see: I declare also to you the signs of joy and rejoycing, even that thing which does fix its Body.

Salmon. That is, he declares the cause of life and death, to be in the Spirits, to wit in the natural Spirits, whether Animal, Vegetable, or Mineral. He who knows how to revive dead Minerals, and to purify them, knows how to exert their powers, and is in the Highway to the greatest of Secrets. 'Tis this Spirit, joyned with its Philosophick Earth, which has power to fix both perfect and imperfect bodies, and to tinge them into the highest perfection of Silver and Gold, which he calls the signs of joy and rejoycing.

VIII. Hermes. Now these things our Ancestors gave us only in Figures and Types, how they attained to the knowledge of this Secret; but behold, they are dead: I have now opened the Riddle, I have demonstrated the proposition so much desired, so much aimed at: I have opened the Book (of Secrets) to the Skilful and Learned, yet I have also a little concealed the hidden Mystery.

Salmon. He declares, that the ancient Philosophers delivered the Matter, and Process of the Philosophick Tincture in Aenigma's and Types & Shadows only; they left no footsteps of the true thing behind them, but what every one might think of at pleasure; therefore from them our Hermes could receive nothing; and he professes, Ch. 1. Sect. 1. That he obtained the knowledge of this Art, by the inspiration of the Living God only; God it was who did reveal and open the Secret to him. This Secret he has opened in this Work, and made so plain, that the skilful and learned may understand it: 'Tis true, he has not unfolded every particular; but yet he has made things so plain, that he who can read him with a Philosophick mind, may at length haply find out the truth: notwithstanding what he has revealed, he declares, he has a little concealed the hidden Mistery.

IX. Hermes. I have kept the things (which out to be put a part) within their own bounds: I conjoyned the various and divers figures and forms (of its appearance in the operation) and I have consederated or joyned together (with them) the Spirit. Receive you this as the gift of God.

Salmon. The meaning of which is, that he has first separated what out to be separated, viz. the pure from the impure, and the Spirit from the Body, which is the first work in order to putrefaction, corruption, and death. Then secondly, he has joyned again what ought

to be conjoyned, to wit, the various and divers
figures and forms, the Soul with the Body, that it may
again be enformed with Tincture and Substance.
Thirdly, he has confederated, or joyned together with
them; the Spirit, which ties the Particles of the Body
and Tincture so firmly together, that they can never
be separated, and unites them in a perpetual
conjunction with a fixity, which will endure for ever.

CHAPTER XI

THE PRACTICAL PART FARTHER EXPLICATED.

I. Hermes. It behoves you therefore to give thanks to God, who has largely given (of his bounty) to all the Wise; who delivers us out of the Snares and Clutches of Misery and Poverty.

Salmon. For this inestimable Gift of God, it is but gratitude to return him the Tributes of Humility and Thanksgiving; to abase our selves before his Divine Majesty, with all humbleness and submission; who thus raises you out of the Dust to fit among Princes, making you to despise the Glories of Crowns and Scepters as insignificant Baubles, and to rest with infinite content in the meanness of a despicable Cottage, for that you carry within your Brests the true Treasure more valuable than all the whole World besides.

II. Hermes. I am proved and tried with the fulness of his Riches and Goddness; with his probable miracles; and I humbly pray God, that whilst I live, I may pass the whole Course of my Life, so as I may attain him.

Salmon. When a Man becomes Master of his Arcanum, he is then tried and proven indeed, how in the midst of such a fulness of Riches and Happiness he can humble himself, and sink in to the deep Abyss of nothingness, abstracting himself from all the goodly things of this life; In this humble state God is only to be met with, (for the proud he beholds afar off) and in this abjection and self-denial, in this mortification of the first life and birth, a second is to be found, a being brought forth in the love of God, the birth of the new Man formed after the Image of the second Adam, a new Spirit, a new Life joyned and United to the Life of God, which can never Perish or Decay, a Fountain of Eternal Delights, an inexhaustible Treasure, infinitely exceeding that which we have all this while thus earnestly been seeking after, and pursuing.

III. Hermes. Take then from thence the Fats or Sulphurour Matter, which we take from Suets, Grease, Hair, Verdigrease, Tragacanth, and Bones, which things are written in the Books of the Ancients.

Salmon. By the Fats or Sulphurous Matter understand, the Sulphurs of all kinds educed by the Aichymick Art, out of Natural things, of which Sulphurs, one only is fixed, and incombustible, and it is a thing which is both in the Earth and in the Heavens; it is in Act,

Animal, Vegetable, and Mineral, found every where, known but by a few, and expressed by its proper Name by no Body, shadowed forth under Various Figures and Aenigmaes. This fixed Sulphur, the Philosophers understand to be nothing else, but the true Balsam of Nature, with which the Dead Bodies of the Metals are imbibed, and as it were throughly moistned, to preserve them perpetually from Corruption. The more any thing abounds with this Balsam, the longer it lives, and is preserved from perishing. From things therefore abounding with a Balsam of this kind; is this Our Universal Medicine drawn; which (as well as for Metals) is made most effectual to conserve Humane Bodies in a State of Health, and to root out all sorts of Diseases, whether accidental after the Birth, or Hereditary by Propagation, restoring the Sick to their pristine Health and Integrity. This Sulphur is not taken from Snets, Grease, Hair, Verdigrease, Tragacanth, Bones, etc. But under these and other the like Names, our Hermes by a Philosophick Liberty, has vailed the Verity from impious and unworthy Men.

IV. Hermes. But the Fats which contain the Tinctures, which coagulate the Fugitive, and set forth, or adorn the Sulphurs , it behoves us to explicate their disposition (more fully hereafter.)

Salmon. Here, in more words, Hermes explicates the Condition, or Qualities, and Properties, of the true Balsam of Nature, or Philosophick Sulphur.

1. He says it contains the Tinctures.
2. It Coagulates Fugitive Substances.
3. It exalts the Power of the Sulphurs, by fixing the Volatile, and making Bright and Shining the things which were Dark and obscure.

The Volatiles of this kind, are nothing else but all the inferior and imperfect Metals, which by this Balsam or Sulphur, are transmutted into the best and finest Silver and Gold. Now this hidden Sulphur dwells in the Bodies, just as Fire in a Coal, or Natural Heat in a Humane Body, or the Vegetative Life in the Spring time, in Herbs, Plants, and Trees, which in Process of time, makes them bring forth Buds, Leaves, Flowers, and at length perfect Ripe Fruits and Seeds. Or like Heat in the inward parts of the Earth, and Bowels of the Mountains, where the most simple Bodies of things, or Elements are first mixed, and produce Metals, Minerals, Stones, etc. according to their several varieties and kinds. So this our Sulphur of Nature contains in it self the true Tinctures, which by the revolution of time it explicates; making ripe the unripe, purifying the impure, fixing the Volatile and ennobling the Ignoble and Vile.

V. Hermes. And to unveil the figure or form, from all other Fats or Sulphurs (which is the Hidden and Buried Fat or Sulphur) which is seen in no disposition, but dwells in its own Body, as fire or heat in Trees and Stones, which by the most subtle Art and Ingenuity it behoves us to extract without Burning.

Salmon. It unveils the Figure or Form, distinguishing itself from all other Fats, Balsams, or Sulphurs; He calls it Hidden and Buried, because it is not Vulgarly known, but only to such as are Adepts: And Buried, because it lies Centrally in the Bodies of Sol, Luna, and Mercury, as a thing Buried in the bowels of the Earth: It is seen in no disposition, but dwells in its own Body, that is, it is not perceptible in any of the imperfect Metals, because they have not Bodies able to hold it, till by it they are made pure and fixt, where-they may become as its own Body is, and so takes up its habitation and dwells in them, as Heat does in Trees in the Spring time, when the External Heat of Sol, stirring up their internal or Mercurial Heat latent within them, makes them bud, and bring forth Leaves, Flowers, Fruits, and Seeds, and that to perfection. This Sulphur (saith Hermes) it behoves us to extract without Burning; for in the Mercury it is yet Volatile, and therefore by subliming of it more

113

1. The Water, which is our Mercury.

2. Gold, which is Sulphur.

3. The mean, or almost Gold, which is Our Salt, or Philosophick Earth, and is more worthy than either the Water or the Foeces, by which Vulgar Gold may by projection be tinged, and made more than perfect.

This is that pretious Stone, in comparison of which, Gold it self, the most pure Gold, is esteemed but as a little Sand, and Silver as Clay in respect thereof. This Gold in a mean, is Gold, in a middle principle, that is, Essential Gold in the Root of the Aurisick Agent, which is in the possibility of augmentation or encrease, even as a very little Plant which becomes a great and mighty Tree; now this third principle which he calls Gold in a mean, is the very Soul it self, which makes this our Philosophick Plant to grow, giving it form and Beauty, and making it become a Golden Tree of a vast and almost infinite magnitude.

VIII. Hermes. And in these three are the Vapors, the Blackness, and the Death.

Salmon. That is in one only Subject composed of three, Spirit, Soul, and Body, these three Vapor, Blackness, and Death are latent, which three are also one. The Caput mortuum must be dissolved; for except the Body

116

be dissolved, there can be no Coagulation of the Spirits: for the Solution of the impure and vaporous Body, induces and brings forth more pure and Noble Spirits, indued with a mighty Strength and Power. And by means of this Solution, a more perfect mixtion is made as of Water with Water, which cannot be separated; not like that of Sand with Sand, whose Superficies only touch one another, which is indeed no true mixtion. And thus by making a dissolution of the Metalline Principle, that which is not Metalline, nor will dissolve, nor mix with the dissolved Matter, (as the Vapor, the Blackness, and the Death or Putrefaction,) comes to be separated and removed, whereby the Dead comes to Live, and that which was in Captivity and Chains comes to be made free, delivered, and set at Liberty.

IX. *Hermes. It behoves us therefore to chase or drive away, and expel the Super existent Fume or Vapor, from the Water;* the *Blackness from* the *Fat; And the Death from the Foeces, and this by Dissolution: By which means we attain to the knowledge of the greatest Philosophy, and the sublime Secret of all Secrets.*

Salmon. In these three, that is, in the One, Composed of the three, lie these other three, the Fume, the Blackness, and the Death, that is, the want of

Ponderosity, of Tincture, and of Fixity, both which threes in their own principles, are also but one thing, to wit, the Caput mortuum, which is depurated and revived by Dissolution only. And except the Body is Dissolved, there can be no Coagulation of the Spirits, as we have said before. And therefore if you would remove the Fumes, you must dissolve the Fumous, or imperfect Body, that it may mix with the Ponderating Spirit. The Fat or Sulphur is cleared from the blackness by manifold Sublimations, bringing forth the pure Philosophick White and Red Flowers, which are the Tincture. And the Death is expelled by the Mercurial or Metallick Spirit, which gives the Eternal fixity.

than Vulgar Silver or Gold, or any Gem, or Pretious Stone. Many have sought this Aeris Viriditatem in Vitriol; and Copper or Vulgar Brass, but they erred and were deceived, following the literal Discourse of the Philosophers, and not their Sense: For they ought not to have contemplated the Metals as they are Bodies, but as they are reduced into a most Subtil, Spirituous, and Celestial Substance.

II. Hermes. Therefore the Philosophers bear up, and magnifie themselves in it, saying that such Gold in Bodies is like the Sun among the Stars, most Light and Splendid. And as by the Power of God, every Vegetable, and all the Fruits of the Earth are perfected; so by the same Power, the Gold, and (the Seed thereof) which contains all these seven Bodies, makes them to spring to be ripened, and brought to perfection, and without which this Work can in no wise be performed.

Salmon. As Sol is among the Stars and other Planets, and Vulgar Gold among the other Vulgar Metals and Minerals; so also is our Gold (which is the true Philosophick Tincture) among the other Metals or Bodys reduced to a Spirituality and pure Tincture: And as Sol in the Heavens is the Medium that perfects all Sublunary or Inferior things by his Beams, Light, and Heat: So also Our Sol, (the true Seed of Gold, and the

121

Seminal Power of the Aurisick Principle) is also the
Medium which makes all the other seven Bodies not only
perfect, but more than perfect; that they thereby may
perfect other quantities of their own kind, yet lying
in imperfection, viz. wanting Purity, Tincture and
Fixation: All which is done by Virtue of its subtle
Spirit, Tincture, and Fire. Therefore say the
Philosophers, Our Gold is not Corporeal, but a
depurated substance in the highest degree, and brought
to an Astral, or Heavenly Nature: This is the Ixir,
Elixir, or Fermentum, the true Tincture and Spirit,
tinging and fixing all other Bodies, and without which
they cannot be perfected.

*III. Hermes. And like as Paste or Dough is impossible
to be Fermented, or Levened without Leven, so is it in
this case, without the proper Ferment, you can do
nothing: When you sublime the Bodies, and Purifie them
separating the filthiness and uncleanness from them,
or from the Foeces, you must conjoyn and mix them
together, and put in the Ferment, making up the Earth
with the Water.*

Salmon. Our Hermes, a little before has made mention
of Ferment, which he has in plain, open and manifest
Words, declared to be Gold: He now comes to
demonstrate the necessity of Fermentation, setting

some of its Operations in Order. The other Imperfect Bodies are the Meal or Dough, and unless they be Fermented with their proper Leven, which is Gold, they cannot be brought into the property of the Leven or Gold: but this Gold must be made spiritual and living, and the Bodies must be Dissolved, Sublimed, and Putref led, before they can be mixed with the Ferment; this being done, viz. being made clean, subtil, and spiritual, the Ferment or prepared Gold is to be mixed therewith, making up the Earth with the Water, that is the Body with the Spirit. Now to bring the Bodies into this State, to be fit to be joyned with the Ferment, you must sublime them, purifie them, make a separation of the Foeces, then conjoyn and mix; all which are necessary in Order to this Fermentation. The Ferment to the prepared Body, is as the Soul to the Body, or as Leaven to Paste, without which the Mass could not be levened.

IV. *Hermes. And you must Decoct and Digest till Ixir, the Ferment, makes the alteration or change, like as Leven does in Paste. Meditate upon this, and see whether the Ferment to this Compositum, does make or change it from its former Nature to another thing. Consider also that there is no Leven or Ferment but from the Paste itself.*

Salmon. Now he teaches us the Art of Levening; which

is to Decoct or Digest, till the Ferment makes an
alteration or change, like as Leven does in Paste.
This is a high point of Art, and ought to be seriously
considered, even what the end of the intention is,
which is to produce or generate Gold; and therefore
(as I said above) Gold must be your Ferment. As Leven
is to Paste, so is this Gold or Ferment to Our
Mercury, which is the prepared Body: And as Leven is
made out of the same matter, out of which the Paste is
made; so this Gold or Ferment is made out of the same
principles, viz. Mercury and Sulphur, which our
prepared Bodies come from; therefore Hermes bids you
consider it, and tells you plainly, that there is no
Leven, or Ferment, but from the Paste it self; and
therefore Our Philosophick Gold, which is Ixir, the
Ferment must be prepared from the Philosophick Mercury
and Sulphur in a fit proportion; that when it Works,
it may purge out the Old Leven with all its effects,
which are uncleanness, want of Tincture, and want of
fixity, and so bring forth a regenerate matter, even a
new substance or body, not according to the Old Leven,
but according to the Nature of the New, which is
wholly purity in the height of Tincture and the
strongest fixity. Now this Fermentum is said sometimes
to be two fold, viz. Fermentum Lapidis Aurisici, which
is from Gold, and Fermentum Lapidis Argentifici, which
is from Silver. This is a weighty thing, and worthy to
be seriously considered of, and therefore advises us
to meditate upon it; except the Paste does receive the

Virtues and Properties of the Leven into itself, it cannot be Levened: If it does, it becomes, by a sufficient Digestion, absolutely the same thing with the Leven, both in its substance and properties, and all other respects.

V.   Hermes. It is also to be noted, that the Ferment does Whiten the Confection or Compositum; and forbids or hinders the Burning: It contains, holds, or fixes the Tincture, so that it cannot fly away, and rejoyces the Bodies, and makes them mutually to joyn, and to enter one into another.

Salmon. He says here, that the Ferment does Whiten the Confection, concerning which Ferment a great doubt does arise, but it is easily solved Philosophically thus. It is not Gold, except it be first Silver. Our Gold is the Tincture, or Soul, or Nourisher of the Work, without which it can never be done: nor is it made Silver, unless it be first Mercury: so that our Sol seems to appear with 3 Faces; first Black, which is the Putrefaction of the Mercury. 2. White, which is the change or transmutation of the black Mercury into a White body, or Silver. 3. Red, which is also the transmutation of the White body, or Silver into a Red Tincture or Gold: so that you may see that this Fermentum not only Whitens the Confection, but also keeps it from Burning, and so fixes the Tincture that it cannot change, vanish, or fly away. By rejoycing the Bodies, he means a replenishing them with a fixed Tincture, and a fixed substance, to wit, the Ingression of the Ferment into them by Projection: but because the Ferment is not able to enter into Dead Bodies, therefore they must be removed, and made Alive by help of the Aqua Medians, or Mediating Water, which is the Aqua Philosophica, which dissolves, subtilizes and spiritualizes them, which makes also a Marriage or Conjunction between the said Ferment, and the White Earth: And in every Fermentation you ought to take notice of the Weight of every thing. If therefore you would Ferment the White Foliated Earth, to the White Elixir, that it may be projected upon bodies diminished from perfection, you must take of the White or Foliated Earth three parts: Of the reserved Aqua Vitae two parts: Of the Ferment half part: Now if you work for the White, you Ferment must be so prepared, that it may be made a White Calx, fixt and subtil: but if for the Red a most pure Yellow or Citrine Calx of Gold.

VI. Hermes. And this is the Key of the Philosophers, and the end of all their Works: And by this Science the Bodies are meliorated, and restored: and the Work of them (Deo annuente) is performed and perfected.

126

Salmon. This Art of Levening or Fermentation is that which he calls the Key of the Philosophers, i.e. the Key which opens the Door into the Secrets and Mysteries of this whole Work: Of so great Virtue and Power is this Work of Fermenting, that he is bold to call it even the Key of the Philosophers: that is the beginning, middle and end of the Work, both for the White and the Red: so that by the Power and Efficacy thereof, the Bodies may be Renovated, and Exalted into a higher State of Perfection, than what they are by Nature.

CHAPTER XIII

THE NATURE OF THE FERMENT FARTHER EXPLICATED.

I.   Hermes. But by Negligence and an ill Opinion of the matter, the Operations may be spoiled and destroyed; as in a Mass of Levened Paste: Or Milk turned with Rennet for Cheese; and Musk among Aromaticks.

Salmon. Without doubt an error may easily be committed in the Work of Fermentation, if you have a false conception thereof, or be ignorant of its Power, whereby you may miss the end; and be frustrate of your Expectations, losing all your cost and time; as is seen in the Levening of Bread; if you trouble the Mass of Meal and Water too soon, it will not be Levened: If it lies too long, it will be over done; so in our Work, if you be too hasty, you will perform nothing at all: If too long, and with too Violent a Fire, you will hazard the breaking of your Vessel, and by an over Volatility, frustrate the fixity of your Medicine. The making of Cheese is Famous, for almost every Housewife can tell you how easie it is to ruin or spoil all, (how good soever your Milk and Rennet may be,) if you be unskilful in the Art: If the Milk be too hot, or too cold, or the Rennet be too much or too little, or the Coagulum lies too short a time, or too long, you may spoil your Cheese, and miss the Perfection, or Goodness, which therein you seek after. These are Familiar examples, and need no farther exposition. The Matter therefore is, first by our Ferment corrupted, and brought into a blackness by Death, but not such a blackness, out of which it cannot be recovered; but so that in the Course of the Fermentation, the Mass of the Confection may pass through the mutation or changes of all the Colors. Now Heat working at the first in humidity brings forth the blackness; but Heat working in the dryness, causeth Whiteness, and in the White the Citrinity and wonderful deep Redness. These Varieties of Colors are caused only by the Ferment in a proper and fit heat, so that the Corruption of one is the Generation of another; and the Ferment becomes the Ferment of the Ferment, as the Philosophers speak. He who cannot taste the Sapor of Salt, will never attain to this desired Ferment of Ferments, which is the Soul, even before Fermentation. If therefore this Ferment be not well prepared, your Magistery will be nothing worth: and know, that this Fermentum is taken only from Sol and Luna, that is, from Gold and Silver, and converts the other Bodies into its own Nature: Therefore it behoves you to know how to introduce this Ferment into Dead and imperfect Bodys (that is, to make Ingression) because it is the Soul; and this Soul gives to them Life and Perfection; so that together with this living and perfect Soul, they are made alive and perfect, and one perfect Body.

II. Hermes. The certain Color of the Golden matter for the Red, and the Nature thereof is not sweetness, there-for of them we make Sericum, which is Ixir, (the Ferment:) and of them we make Enamel, of which we have Written.

Salmon. Altho' it does not here sufficiently appear what our Author means by Sweetness and Sericum, yet afterwards he so explains himself that we may guess at it; and that it is the Golden Ferment for the Red; the adumbration whereof he gives us under the Mask of Encaustum or Enamel; and truly by Figures, Similitude, and Tropical ways of speaking, he has been pleased to deliver himself through this whole Work. I suppose he uses the Similitude of Sweetness here in respect of Leven; for that Leven is not Sweet. *(Sericon: See Ripley-Bosom Book -HWN)*

III. Hermes. And with the King's Seal we have tinged the Clay, and in that we have put or placed the color of Heaven, which augments the sight of them, who can already in some measure see.

Salmon. By the King's Seal is meant the Virtue, Power,

Character, or Tincture of Gold, which tinges Lutum the Clay, that is, the Mercurial Mass, or Earth, which is now but one thing, and a Secret drawn out of the Fountains of the Wise, for which reason it is be some called Sigillum Sapientum: Also Sigillum Hermetis, and Sigillum Mercurii. This is the thing which may have sought after in vain, and could never find, that is, the outward turned inward, and the inward parts turned outwards; that which was below raised up, and that which was above, laid down below; the Superiors and Inferiors, the Heavens and the Earth joyned together in one Globe or Mass, and digested together in one, till they produce the heavenly color, the light of Sol, which gives such as have Eyes to see, the happiness of seeing a Fountain inexhaustible, an Eternal Spring, the permanent and endless Treasure.

IV. Hermes. Gold therefore is the most pretious Stone without Spots, also temperate, which neither Fire, nor Air, nor Water, nor Earth, is able to corrupt or destroy, the universal ferment, rectifying all things, in a middle or temperate Composition, which is of a Yellow, or true Citrine colour.

Salmon. Our Hermes here confesses plainly, that the Philosophick Gold, is this most pretious Stone, without blemish and incorruptible, and differs as much

132

from vulgar Gold, as Leven does from the Paste, or Yest from the Ale or Beer which is made by it: For as clear, well—wrought Ale cannot change other Wort into Ale, nor Levened Paste leven another Mass of Meal and Water, (till it is brought to the perfection of Leven) so neither can vulgar Gold (which is the product of Mercury and Sulphur) transmute, or change any other body into its own Purity, Tincture, and Fixity. No: This is only the work of our Stone, Elixir, Tincture, the true Philosophick Gold.

V. Hermes. The Gold of the Wise Men, boiled and well digested, with a fiery Water makes Ixir.

Salmon. The Gold is to be exquisitely boiled, as much as you please with a fiery water, and digested: This fire is found no where more perfect, better, or more powerful than in Minerals and their Roots, which Roots the Philosophers say, are in the Air: And the Gold is Spiritual Gold, not the body of vulgar Gold unprepared. This Aqua Ignea, is nothing else, but the Mercury of the Philosophers, drawn from its Mineral Root. This Water is the Mother, which does dissolve the Gold conceived in its Belly, being digested and nourished there for forty Weeks, at the end of which digestion, like as in the hour of a mans Nativity, the Soul (i.e. the Tincture arises) but not first nor quickly. In this point is all the hazard; but this being past, there is no more peril, the danger is wholly over.

VI. Hermes. For the Gold of the Wise Men is more weighty or heavy than Lead, which in a Temperate (or due) composition, is the ferment of Ixir: and contrariwise, in a distemperate (or undue) Composition; the distemperature, or hurt of the whole Work or Matter.

Salmon. Our Gold, the Off-spring of this great Work, is much heavier than Lead, because of its Weakness, Volatility, and Intemperature: Our Infant is of a most strong and temperate Composition, healing the Infirmities of its proper Parents, and tinging the Mercury of all Bodies whatsoever, into the best and most pure fine Gold. By this is understood the Vital Roots of the Minerals, into which, if the Bodies be reduced, they are made apt, or fit for a new Regeneration, so that from the same you may have the true Tincture of the Philosophers.

VII. Hermes. For the work is first made from the Vege-table: Secondly from the Animal, in a Hens Egg; in which is the greatest assistance, and the constancy of the Elements. And Gold is our Earth; of all which, we make Sericum, which is our Ferment, or Ixir.

Salmon. He here divides the great Work into two parts, viz. Vegetable and Animal, which is a Philosophical fiction: But the true Work is but one, consisting of an equal and temperate mixtion of the Elements, to a perfect fixity. The Foundation of this Work, is laid in the Earth of the Gold, of which the Ixir, Elixir, or Ferment is made, which is two fold. 1. For Luna. 2. For Sol. By the Ferment of Sol is understood the Seed of the Male: and by that of Luna, the Seed of the Female: of these there must be made, 1. A Conjunction. 2. A Generation. The Ferment of Sol, is from Sol; as Leaven is made of the substance of the Bread; and as a little Leaven, Ferments, or Leavens a great quantity of Paste, (at least 250 times its quantity;) so likewise a little Portion of this our Earth suffices to nourish and perfect the whole Stone. The Ferment, saith Avicenna, reduces the Matter to its own Nature, Color, Sapor, and Form, reducing Power into Act. For it Whitens the Confection, Multiplies it, makes it Spiritual, Strengthens it, makes it resist the Fire, makes it contain the Tincture, that it shall not fly away, opens the Bodies and makes them, with it, to enter one into another, and to be perfectly conjoyned, as Water with Water, which cannot be separated, and is the end of the Work. Without this Ferment, no Elixir can be made, no more than Paste or Dough can be Levened without Leven. And this Elixir is the Ferment of Ferments and the Coagulum of the Coagulum. For, it not only Ferments the Inferior and imperfect Bodies, but also Gold it self; making it from a perfect Body, much more than perfect. It is the most faithful Mother, which by how much the oftener it is impregnated, by so much the more it conceives and brings forth, propagating its Off-Spring to an Infinity of Generations. It is the only Key which opens and shuts the Gates leading to the Kingdom of the Mineral Treasure, the Golden Mountain, the Gardens of the Hesperides, where all the Trees perpetually bear Golden Fruit. Without this Key, it is not possible for any Man to attain to the perfection of this Art.

CHAPTER XIV

THE SMARAGDINE TABLE OF HERMES.

I.   Hermes. This is true, and far distant from a Lie; whatsoever is below, is like that which is above; and that which is above, is like that which is below: By this are acquired and perfected the Miracles of the One Thing.

Salmon. That is to say, the truth of this our Art is confirmed by Experience, we know it to be truth by very matter of Fact; and notwithstanding all the Sophisms, and Logomachia of the Schools, there is no Argument can stand against Experience. The Waters of the Cataracts of Heaven above, are like to the Waters below, when the great Fountain of the deep is broken up; and the Waters below, are like to the Waters above. There are two parts in our Stone, a Superior part that ascends up, and an Inferior part which remains beneath; and yet these two parts agree in One. The inferior Part or Earth, is called the Body or Ferment. The superior, part or Spirit, is called the Soul or Life, which quickens the Stone, and raises it up: The first must be dissolved, and made Water, like the Superior; and the Superior must be coagulated, and made Earth, like the Lower, that they may be united, and become the Miracle of the one Thing; then will it be evidently demonstrated, that whatsoever is below, is like that which is above, and contrarywise. Nor do they differ one from another but by Accident, as Corruptible and Incorruptible, Pure and Impure, Heavy and Light, Clear and Opake, Agent and Patient, Masculine and Feminine, etc. all which are Accidents, not Substances. Heaven or that which is above is Incorruptible, where the pure Elements are made, from a Corruptible matter elevated or lifted up, in the Concavity of which Firmament, the Body or Substance of Luna is Graduated. Hence it is apparent that this our Medicine must resemble Heaven it self, in Activity, Penetrativeness, and Incorruptibility; nor must it work as the Elements in Natural Bodies, which are as it were Dead, and destitute of any Power or Action.

II. *Hermes. Also, as all things were made from One, by* the *help of One: So all things are made from One thing by Conjunction.*

Salmon. That is, as all things were made or came from One Confused Chaos, by the help of One Omnipotent or Almighty God; so Our Stone is born or brought fourth out of one Confused Mass, by the help of one particular Matter or Thing, which contains in it four Elements, Created by the determination of God. Here Hermes points forth the Universal Medicine in

138

imitation of the Worlds Creation; which is performed by one Universal Spirit, and so by a Supernatural Experiment, points forth this Our Natural Work. It is the Opinion of many Philosophers that the Spirit of Natural things, or the Spirit of Nature is the Medium between the Soul and the Body, as being that which makes the absolute and firm Conjunction. But the Opinion of some is though the Spirit may be said to be the more subtile Substance; yet it can be no more separated from the Soul, than Light from the Sun.

III. Hermes. The Father thereof is the Sun, and the Mother thereof is the Moon: the Wind carries it in its Belly, and the Nurse thereof is the Earth.

Salmon. As living Creatures beget their Like or Kind, so Gold generates Gold by the Virtue of Our Stone. The Sun is its Father, that is, Our Philosophical or Living Gold. And as in every natural Generation, there must be a fit and convenient receptacle, with a certain likeness of kind to the Father; so likewise in this Our Artificial Generation, it is requisite that the Sun, or Our Living Gold, should have a fit and agreeable Receptacle or Womb, for its Seed or Tincture; and this is Our Philosophical or Living Silver, i.e. Mercury, which is the Mother thereof. What Sol and Luna are in the Heavens above, the same are Our Gold and Silver in Our Heavens below. The Universal Masculine Seed is the Sulphur of Nature, the first and most Potent cause of all Generation: And if Sol does Live, it is necessary, as Paracelsus saith, to live in some things, viz. in its own Radical Humidity, and most pure and simple Air, which contemperates the heat thereof by its Humidity. The Wind is the Air, and the Air is the Life, and the Life is the Soul, which quickens the whole Stone. And therefore the Wind, Air, Life or Soul must carry the Stone, viz. bring forth Our Magistery: which being brought forth, it must be nourished by its Nurse, which is the Earth; for The Earth (saith Hermes) is its Nurse. The Wind Carries it in its Belly; by which the Universal, Inferior, and Feminine Seed is dilated through the Air, and joyned to the Universal Superior and Masculine Seed; the Air or is the Womb wherein the two Seeds are conjoyned. The Air arises from Fire and Water, as the Heaven from Fire and Air. Under the Appelation of Fire, is comprehended the most pure substance of the Earth, ascending with Fire: and under the Name of Air the most pure Substance of Water; the Belly or Womb of Nature, is a most pure Breath or Matter, raised from all the inferior Elements, converted into a Volatility or Air, in which is conceived by the help of Luna, the Universal Seed of the Sun, specificated also by the other Lights or Stars. Hermes will have three Elements, two under the Names of Sol and Luna, the third under the Name of

Ventus, the Wind. The Earth is the Nurse of this Birth of the Air, by whose Breasts it is Nourished, whence it Sucks the Mercurial Milk, (that is the more thick substance of the Inferior Water remaining yet in the Earth) by which it grows and increases to its Substance and Perfection, as a Child to the Stature and Strength of a Man.

IV. *Hermes. This is the Mother or Fountain of all Perfection, and its Power is Perfect and Intire, if it be changed into Earth.*

Salmon. As if he should say, this Arcantum which I here shew you is the Original and Fountain of all Arcanums and Mysteries, the secret Treasure of the whole World. But it is not brought to its Perfection till it is changed into Earth; then indeed is its Power perfect and intire: that is, if the Soul of the Stone (of which we have spoken before and which may be called the Wind or Air, which is also the Life, the Virtue, Power and Spirit) be converted into Earth, viz. a fixt Substance or Matter; so that the whole Air, Spirit, Life and Soul of the Stone may be conjoyned to its Earth, which is its Nurse, and be all turned into Ferment. As in making of Bread, a little Leaven Ferments and Transmutes, a great deal of Meal or Paste: so also must Our Stone be Fermented, that it

may become Ferment to the Eternal Multiplication
thereof. That which the Wind does bear in its Belly
must be converted into Earth, then is the Work
compleated; which is done by a long and Unwearied
Decoction (not by evaporating, but retaining the
Spirits) till it becomes inspissated, and in success
of time is dryed into a Pouder or Earth. But the time
will be long and tedious, therefore you must attend it
with Patience, according to the Matter you work upon.
Some things are remote from Perfection, other things
more remote, and others most remote, whilst other
things are near to Perfection, others neerer, and some
things nearest of all. He that knows not these things
before he begins his Work, may afterwards deplore his
Error, with very great loss.

V.   Hermes. *Separate the Earth from the Fire, and the*
Subtile and Thin from the Gross and Thick; but
prudently with long Sufferance, Gentleness and
Patience, and Wisdom, and Judgment.

Salmon. Hetherto he has only discoursed the Theory, he
now comes to shew you the Practical part, shewing
first the Purification of the Matter of the Stone. You
must do it gently, by little and little, not
Violently, but Prudently and Wisely, after a
Philosophick manner: By Separating he means

Dissolving: for Dissolution is the Separation of parts: Some will have it, that by the Earth here, he means the Lees or Dreggs of the Matter, which is to be Separated from the Fire, the Air, and the Water, and the whole Substance of the Stone, that it may become Pure, and free from any Putrefaction or Defiled Matter: and this Spagyrick Philosophers say is the first Operation or Preparation of the Matter or Parts of their Stone. But some understand hereby, the Separation of the four Elements, and this doubtless is the thing if it be spoken of a Spagyrick, and not Vulgar Separation. Under the Appellation of Fire, the two other are understood, viz. Air and Water; for the Fire cannot want to subsist without Air, nor is the Air without Water; for Air is made of Water by the Mediation of the Fire, by which it is forced to Ascend upwards. But as to the Earth, it partly Ascends and is made Volatile, and partly remains fixed below. By separating the Earth from the Fire, some will have it, that he would have the Thick to be separated from the Thin, not the Thin from the Thick, because the Earth is thicker than the Fire. But by separating the subtil from the gross, is to be understood, the subtilizing of the Thick Matter, and Spagyrically to reduce that subtilized Matter into Aether or Spiritual Air. But this must prudently be done, with gentleness, long Suffering, Patience etc. that is according to the Laws of Art, but gently, even with a gentle Heat according to Natural Generation. The Instrument of Nature, and of the Spagyrists Fire, without which the Work cannot be done. This Fire is either Internal or External. The Internal is proper to the substance or Matter, and Naturally dwells within it, which you must prudently stir up and Excite. The External is either Violent, or Temperated in four several Degrees. The Violent is that with which some things are Calcined, other Sublimed, others (as Metalls) Liquified or Melted. The Temperate in its several Degrees, imitate or resemble Nature, and are used for Putrefaction, Digestion, and Congelation or Circulatorily to dissolve and fix. But Various are these kinds of Fire, which are to be applyed according as the Subject requires, and the Prudence of the Artist directs, being continual without interruption from beginning to the End.

VI. Hermes. It Ascends from the Earth up to Heaven, and Descends again from the Heaven *to the Earth, and receives the Powers and Efficacy of the Superiors and Inferiors.*

Salmon. Here is to be observed that though Our Stone be divided in the first Operation into four Parts, which are the four Elements, yet as we have already said there is but two principal Parts of it; One which Ascends upwards and is Volatile, and another which remains below, and is fixed, which is called Earth, and ferment which Nourishes and Ferments the whole Stone. But of the unfixed or Volatile part we must have a great quantity, that it may Nourish the purified Matter of the Stone, till it be made to Ascend, is sublimed, and subtilized: then being thus subtilized and made Volatile, it must be incerated with the Oyl, extracted from it in the first Operation, which is called the Water of the Stone, and so often Boyled by Sublimation, till by Virtue of the Fermentation of the Earth exalted with it; the whole Stone again does descend, from the Heaven to the Earth, and remains fixed and flowing; (that is that the Corporeal be made Spiritual by Sublimation, and the Spiritual be made Corporeal by Descension: Here is a Circulatory Distillation admirable declared, and the Construction of a Spagyrical Vessel, to the Similitude of Nature. It Ascends from the Earth, i.e. from the inferior part of the vessel; to Heaven, i.e. the superiour part: The matter generated of Sol, and Luna ascends, i.e. the thick Terrean substance thereof is converted or resolved, into Heaven, viz. into a subtile substance like to Heaven: he demonstrates the Spagyrick solution, by what Instrument and Artifice it is done; then he teaches the Fixation, It Descends again to the Earth; as if he should say, after its substance is dissolved and made to Ascend under the Obedience of the Internal Celestial Virtues or Powers, standing there the determined time of its Maturity, it returns again, or descends, that is to say, the Spirit

is made Corporeal, which was before a Body or made
from a Body, Spiritual, which is nothing but the
Philosophick Riddle. Fac Fixum Volatile, & rurfus
Volatile fixum, & totum Babebis Magisterium. And by
this means it will obtain the Virtues of the Superiour
and Inferiour Powers, i.e. the Heavenly and Volatile
Power, to penetrate, grow, increase or multiply: and
the Earthly Power to give Substance, Corporeity, and
Fixity.

VII. Hermes. In this Work, you acquire to yourself the
Wealth and Glory of the whole World: Drive therefore
from you all Cloudiness or Obscurity, Darkness and
Blindness.

Salmon. Possessing this Stone thus perfected, you
possess all the Wealth and Treasures of the World; so
that you may live free from Care and Trouble, from Dis-
content and Fears, from every Sickness and Disease:
It is a Remedy for all Diseases both of Body and Mind:
It strikes at the root of Infirmities; and destroyes
that which would destroy or undermine the Health and
Prosperity of the Humane Body. This Stone, this
Wealth, this Treasure, though it be but like to a
Grain of Mustardseed, yet it grows to be the greatest
of all Trees, in whose Branches the Birds of the Air
make their Nests, and under whose shadow the Beasts of

the Field dwell.

VIII. Hermes. For the *Work increasing or going on in Strength, adds Strength to Strength, forestalling and over-topping all other Fortitudes and Powers; and is able to Subjugate and Conquer all things, whether they be thin and Subtil, or thick and Solid Bodies.*

Salmon. There is no Comparison of the Powers of other Natural things, to the Power of the Stone, for it is able to overcome and master all other Powers: it can convert common Quick Silver into a Congealed substance, and Transmute it into fine Gold or Silver: and it can Penetrate and Peirce through all other hard solid or compact Bodies, and strike them with a never fading Tincture, so firm and fixt, which the Power and Strength of the Strongest and most Violent Fire can never conquer or overcome. This is as much as if he should say, it is the compleat Virtue of total Nature, the Power, Efficacy and Potency of all things, and even (as it were) above Nature, so that it may not improperly be said to be a Work Metaphysical, for that it seems to act above or beyond Nature. It overcomes or conquers all things, that is, it makes all subtil and thin things (as QuickSilver) thick and coagulates them: and on the contrary it Penetrates all thick and solid things, i.e. It makes every hard Metal whether

Perfect or Imperfect (as Sol, Luna, Saturn, Jupiter, Mars and Venus) subtile and thin, and brings them to the greatest Perfection, expelling all the Malign and Dark Spirits possessing them and giving to them Tincture and Fixity, by its Subtility and Spiritualty.

IX. Hermes. In this manner was the World made; and hence are the wonderful Conjunctions or Joynings together of the Matter and Parts thereof, and the Marvellous Effects, when in this way it is done, by which these Wonders are Effected.

Salmon. The Creation of the World he brings as a Prior Example, or Exemplification of the Work of Our Philoso-phick Stone, for as the World was Created; so is Our Stone composed. As in the beginning the whole World and all that is therein was a Chaos or confused Mass, but afterwards by the Virtue Word, Power, or Spirit of the Great Creator, a Separation was made, the Elements were divided and rectified, and the Universal World was produced and brought forth Beautiful and Perfect in Number, Weight and Measure. So also in this our work, we separate the Elements, which we divide and rectifie by many sublimations, depressions, and precipitations, whereby the perpetual and wonderful conjunction is made, which is the product of the prime matter, and the root of the Golden Kingdom, in which power is produced into Act.

X. Hermes. And for this Cause I am called Hermes Trismegistus, for that I have the knowledge or understanding of the Philosophy of the three principles of the Universe. My Doctrine or Discourse, which I have here delivered concerning this solar Work, is compleat and perfect.

Salmon. Hermes Trismegistus signifies the Thrice greatest Hermes, for that he had the knowledge of the three Principals of the Universe, viz. Salt, Sulphur, and Mercury, answering to the Body, Soul and Spirit; Mineral, Vegetable, and Animal, of which he had the true Knowledge, he knew the way how to separate them, and conjoyn them again, to make the fixt Volatile and the Volatile fixt, to take away Tinctures, and restore better again, all which are contained in Our Philosophick Mercury which is the Womb in which Our Philosophick (which is the true) Gold is Generated. It is said to be perfect, because

1. It contains all the Principles.
2. From its never fading Coler.
3. Its never perishing Body.

It is resembled to a grain of Wheat, which unless it Dies, it brings forth no Fruit; but if it Die, and is Putrefied, passing through Death and Putrefaction or

Dissolution, to Life and Heaven, there by perfecting its Nature, it is infinitely profitable. What he has delivered concerning this Matter, viz. of the three Colers, Black, White, and Red; of the three Principles, Salt, Sulphur, and Mercury; of the three Subsistencies, Body, Soul and Spirit; of the three Operations, Volatilization, Tincture and Fixation; of the three States, Imperfection, Anihilation, and Perfection, he declares to be True and Compleat, and that the Stone thus Generated (existing and being in one only thing, viz. the Philosophick Mercury) by a series of Natural Operations, is Perfect and Intire, wanting nothing.

LIBRI HERMETIS PRIMI.

F I N I S.

CHAPTER XV

THE ENTERANCE INTO THE WORK, BEGINNING WITH ARGENT VIVE.

I. Hermes the Wise saith, if you Conserve a third part of your Camels, (viz. of the swift or Volatile Matter, or that which must bear the Burthen,) and Consume the remaining two thirds, you have attained to the thing desired; you have perfected the Work.

II. In like manner you must be careful of your Argent Vive; for the black Matter does Whiten the Flesh, and the Work is perfected by the Fire of the Wise.

III. And the Work is to be performed by a Spiritual Water, in which the blackness is washed away; and by that Instrument, in which the Foundation of the Work is laid, and in that time and moment, in which the Clouds appear.

IV. Now that Water, in or by which the blackness is washed away, is the Sweat or Moisture of the Sun, and Childrens Urine, (i.e. the Virgins Water.) The thing which I tell you is sufficient for you to know.

V. In like manner, take the Water of the Water, (Mercury of Mercury,) and with it cleanse and purifie the Wind, Fume, or Vapor, and Abolish the blackness. Understand what this signifies, and rejoyce therein.

VI. Also in the same manner, take the blackness and Conjoyn it; then have respect to the White, and Conjoyn the Red; so will you go through the thing desired, and come to the end of the Work.

VII. It is also to be noted, that it is the Fire-Stone which Governs the Matter or Work, by the good pleasure of God: Boyl it therefore with a gentle Fire, Night and Day, lest the Water should be separated from it; even till it becomes of a Golden Color: Understand well what I say.

VIII. That also which Congeals, does Dissolve; and that which does Whiten, does in like manner make Red.

IX. I have made plain to you the nearest way, that you may be easie and satisfied: Understand therefore these things, and Meditate upon them; and you shall certainly attain to the perfection of the Work.

X. It is also to be noted, that as Sol is among the Stars; so is Gold among the other Metalline Bodies: For as the Light of the Sun, is joyned to the Lights, and contains the Fruit of this Operation; so in like manner Gold: Meditate upon these Words, and by the Permission of God you may find it out.

CHAPTER XVI

THE NATURE OF THE MEDICINE, AND GOVERNMENT OF THE METALS.

I. Hermes Moreover saith, he that outwardly takes in this Medicine, it Kills him: but he that inwardly drinks it in, it makes him to Live and rejoyce. Understand what this means.

II. And as for this cause sake, this Water is said to be Divine, so it is said to be the greatest Poyson: and it is preferred before all other things, by so much as that without it nothing of the Work can be done.

III. It is also called Divine, for that it cannot be mixed or joyned with any filth or defiled thing: and this water of our Stone purifies and cleanses the Natures of the Metals, and washes away their defoedations or defilements.

IV. And as Sol Acts upon Bodies, so also does this Water upon the Philosophick Stone: Yea, it penetrates and sinks through it, and is constant, fixed and perfect.

V. This indeed is seen in Sol; but it is to be understood that the Work may be made through all the seven planets: as first from Saturn, then from Jupiter, Mars, Venus, Mercury, and lastly from Luna.

VI. The first is the government of Saturn; to wit, to cause Sol to putrefie, or bring the Body to putrefaction, which is done in the space of 40 days and nights. The second is the government of Jupiter, which is to grind or break the matter, and in 12 days and nights to Imbue or moisten it, which is called the Regiment of Tin.

VII. The third is the government of Mars, which is to induce Death or blackness, or to separate the Spirit from the Body, by which it is said to be changed. The fourth is the government of Sol, which is to Work away the Blackness and Poyson; and is indeed to make it White.

VIII. The fifth is the government of Venus, which is to joyn the moist to the dry, and the hot to the cold, and to Unite them together in one. This is the Dominion of Brass or Copper, (or the making of the matter of a changeable Yellow.) The sixth is the government of Mercury; which is to burn, and is called the Dominion of Argent Vive.

IX. The seventh is the government of Luna, which is to Decoct or Boyl, and make Hot, and so to perfect the matter, (with the fixed Citrine Tincture) in 25 days; and this is the Dominion of Silver. See here, I have gone with you through the whole Work; take heed therefore lest you err.

CHAPTER XVII

THE DIFFERENCE OF THE FERMENTS, AND QUALITY OF THE SPIRIT.

I. And know that the White Body is made with the Whiteness; and its Ferment is that which you already know: Whiten therefore the Body, and understand what I say.

II. Also in like manner you are to note; that the Stone sought after, has not its like or equal in the whole Earth. It is both outwardly and inwardly of a Citrine Golden Color; but when it is altered or changed, it is made a Body dark and black, like burnt Coals.

III. Now the Color of the Spirit taken from it is White, and the substance thereof is Liquid as Water; but the Color of the Soul thereof is Red.

IV. But the Soul and the Spirit thereof is returned to it again, and it doth Live and Rejoyce, and its Light and Glory returns again; and you shall see it overcome and Triumph: And that which was even now Dead, shall have Conquered Death, and then it shall Live, and arise from the Dead, and Live as it were for ever.

V. Happy and Blessed therefore is he in whose Power the disposition of this Matter is, who Kills and makes Alive, and is Omnipotent over all for ever.

VI. I therefore advise you, not to do any thing in this work, till you get an understanding thereof: For if you be Ignorant and void of true Knowledge, you will err in whatsoever you do, you will wholly Labour in Vain, and your Work will Perish.

VII. So that thus mistaking in your Operation, you blame presently your instructors (the Philosophers) and think that they have erred, or taught you wrong, when it is only your Ignorance, and none understanding of their words.

VIII. This then know and understand, that the Day, is the Nativity or bringing forth of the Light; but the Night, the Nativity or bringing forth of the Darkness.

IX. Sol also is the Light of the Day; Luna is the light of the Night; which God Created to govern the World.

X. But Luna does receive her Light of the Sun by Combustion, and is dilated or enlarged therewith: and by so much as she receives of the Light of the Sun, or does contain in her, of his Light; by so much does the Nature of Sol bear Rule over the Nature of Luna.

XI. If therefore you contemplate what I say, and Meditate upon my Words, you will find that I have spoken the Truth; and you will understand the signification, of all that I have said, and the demonstration of the whole Matter.

XII. Know then, that the Spirit, is enfolded or circumscribed, within (as it were) its Marble House or Walls: Open therefore the Passages that the Dead Spirit may go out, and be cast forth from our Bodies: then it will become beautiful, which is only a Work or undertaking of Wisdom.

XIII. Sow therefore (O God) Thy Wisdom in our Hearts, and Root out the corrupt Principles which lodge therein, and leads us in the way of thy Saints, by which our Spirits and Souls may be Purified. Thou art Omnipotent, O Lord God Almighty, and canst do whatever thou pleasest.

CHAPTER XVIII

OF ARGENT VIVE, TINCTURE, ORDER OF THE OPERATION, AND OF
THE FIRE.

I. There is one thing which is to be wondered at, viz. after what manner Carmine, to wit, Grana Nostra, doth tinge or Dye Silk, which is of a contrary Nature, and tinges not a Dead thing: and after what manner Uzisur, to wit, Our Vermilion doth tinge Vestem which is of a contrary Nature, and tinges not Live or growing things.

II. For it is not Natural for anything to tinge other substances, not agreeable to their Natures: If therefore you put into your Composition, Red Gold, you shall find in the Tincture a pure and perfect Red: and if you put into it White Gold; the most passive Red will vanish or go forth. There is nothing indeed does tinge any thing, but what is Consimilar to, or like it self.

III. And I testifie to you by the Living God, maker of Heaven and Earth, that the Stone which I have described, you have permanent or fixed, nor are you kept from it by the Earth or the Sea, or any other matter.

IV. Keep then your congealed Quick-Silver, many parts of which are lost because of its subtilty. Also the Mountain in which is the Tabernacle which crys out; I am the Black of the White, and the White of the Black; I speak the Truth indeed, and I lye not.

V. Now know, that the Root of the Matter is, the Head of the Crow flying without Wings, in the dark and black of the Night, and in the appearance of the Day: from the Throat the fixing Spirit: from its Gall the Coloring or Tinging Matter is taken, from its Tail, the defication, or drying of the Matter; from its Wings the liquid Water; and from its Body the Redness.

VI. Understand the meaning of the words, for hereby is understood our venerable Stone, and the Fume, or Vapour thereof which is exalted (lifted up or sublimed) and the Sea eradicated, and a Light shining.

VII. You are also to note what Alums and Salts are, which flow from Bodies: if you put the Medicines (or Matters of the Medicine) in a just or true proportion, you shall not fear to err; but if you mistake the proportion, you must add or diminish, according as you see it tends to the emendation, or performing of the Work, lest a Deluge should come and overflow all, drowning the Regions, and overturning the Trees by the Roots.

VIII. And tho' the Matter be unknown, yet consider these things, how, or after what manner these two Colors are distinguished, or diversified, by their Vapours: look into the sweetness of Sugar, which is one kind of sweet Juice; and into the sweetness of Honey, which is yet more intense or inward.

IX. Except you make the Bodies spiritual and impalpable, you know not how to putrefie Ixir, or proceed on in the Work; nor how the three Volatile Matters or Principles, fight one against another; and how they fail not, each in their turns, to devour one another, till of two being left, one, only remains.

X. Be careful also, how you increase your Fire (tho it is not to be very small when you dry up the Water) and take heed that you burn not the Matter, because if the Vessel breaks, it will be with a mighty impetus or force.

XI. And unless the Matter of the Stone, prove inimical one to another, or contend and fight with, and strive to destroy one another, you shall never attain to the thing you seek after.

XII. If you mix your Calx with Auripigment, and not in a mean or due proportion, the splendor and glory of the Operation will not succeed; but if you interpose a medium, the effect will immediately follow.

XIII. Now know, that it is our Water, which extracts the hidden Tincture. Behold the Example and understand it; if you have once brought the Body into Ashes, you have operated rightly.

XIV. And the blood (which is in the Philosophick Water) of the animated Body, is the Earth of the Wise, to wit, the permanent or enduring perfection.

CHAPTER XIX

THAT THE BEGINNING OF THIS WORK IS IN THE BLACKNESS AND
DARKNESS: AND OF CONJOYNING THE BODY WITH THE SOUL.

I. Now it is that which is dead, which you ought to vivifie or makes alive; and that which is sick, which ought to be cured: It is the White which is to be rubified; the Black which is to be purified; and the Cold which is to be made hot.

II. It is God himself who does create, and inspire or give life, and replenishes Nature with his Power, that it might follow and imitate his Wisdom, and act as an Instrument subservient to him.

III. Iron is our Gold; and Brass or Copper is our Tincture; Argent Vive is our Glory; Tin is our Silver; Blackness is our Whiteness; and the Whiteness if our Redness.

IV. From hence it appears necessary, that we should have a Body purifying Bodies; and a Water subliming Water. Our Stone which is a Vessel of Fire, is made of Fire; and is converted into the same again.

V. And if you would walk in the true way, you must persue it in the evident or visible Blackness: for (saith our Stone) it is that which is hidden within, which does make me white; and the same thing which makes me White, makes me also Red.

VI. Conceal this thing from Men, like as a word which is yet in thy Mouth, which no Man understands; and as the Fire, Light, or Sight, which is in thine Eye: I will not tell it plainly to thee thy self, lest by thy words thou conveyest my Breath to another, to thine own damage: **This is the caution I give thee.**

VII. Now know that this our Work, is made (or compounded) of two Figures (or Substances) the one of which wants the White Rust (Ceruse) and the other the Redish Rust (Crocus) Our Matters also are Searsed thro our Sieves or Searses, made of pure or clean Rinds, and a most blessed Wood.

VIII. You are also to take notice, that the Fire-Stone of the Philosophers sought after, wants Extention, but it has quantity. It behoves you therefore, to support and nourish it on every side, and to continue it as in

168

the middle.

IX. You must also conjoyn the Body with the Soul, beating and grinding it in the Sun, and imbuing it with the Stone; then puting it into the Fire, so long till all its Stains and Defilements are taken away; let it be a gentle Fire for about seven hours space; thus will you get that which will make you to live.

X. I also tell you, that its habitation or dwelling place, is posited in the Bowels of the Earth, for without Earth it cannot be perfected; Also, its habitation is posited in the Bowels of the Fire, nor without Fire can it be perfected, which is the perfection of our Art.

XI. Again, Except you mix with the White the Red, and presently bring or reduce the same into a perfect Water, it will tinge nothing; for it never tinges any thing Red, but that which is White: and while the Work is now perfecting, add them to the light of the Sun, and it will be compleated Regimine Marino, as we have already declared: and by this conjunction above, your Stone will attain to its Beauty and Glory.

XII. Thus have you a dry Fire which does tinge: an Air or Vapour, which fixes and chains the Volatile Matter, binding the Fugitive in Setters; and also whitens expelling the blackness from Bodies; and a fixed Earth, also receiving the Tincture.

CHAPTER XX

THE ORDER OF THE PRACTICAL PART OF THE OPERATION.

I. Wash your Mercury with the Water of the Sea, till you have taken away from it all its Blackness, so will you accomplish your work to perfection, in which rejoyce.

II. If you understand how a Resurrection is accomplished, i.e. how the living (principle or Spirit) comes forth from the dead (Matter or Substance) how that is made apparent which was hidden, and how Strength is drawn forth through Weakness; you cannot be Ignorant how to compleat and perfect this Work.

III. How Manifest and Clear are the Words of the Wise, yet so as the internal Life and Principal is still hidden; you understand them now perfectly by their Expressions.

IV. Two Bodies equally taken from the Earth, grind, in the Oyl of the Decocted Matter, and in the Milk of the White Volatile: Now mighty and wonderful are the powers and force of these Bodies, which are freely bestowed upon you, through this whole Science, which

you shall possess, and therewith a long and enduring
Life.

V. Take by force the most Intense Wisdom, from whence
you must draw forth the External (perpetual or fixed)
Life of the Stone, till your Stone is congealed, and
its dulness is vanished; so will you accomplish the
Life thereof sought after.

VI. Give therefore of this Life sufficiently to your
Matter, and it will mortifie it (or bring it to putrefaction) but repleat your Earth, and it shall make it
to live (Spring, Bud, Grow, Germinate.)

VII. Plant this Tree upon your Stone, that it may not
be in danger of the violence of Winds, that the
Volatile Influences or Bird of Heaven, may fall upon
it, and by virtue thereof, its Branches may bring
forth much Fruit; from thence Wisdom does arise.

VIII. Take this Volatile Bird, cut off its Head with a
fiery Sword, then Strip it of its Feathers or Wings,
undo its Joynts, and boil it upon Coals, till it is
made, or becomes of one only Color.

IX. Then put the Venom, or Poyson to it, so much as is enough to bury or cover it; govern it now with a gentle Fire, till your Matter is mortified or putrefied; which done, grind it with White Water, and manage it rightly.

X. For we bought two Black _____ (Crows) and we put them into a Paropsidem, or Crucible (or Cupel) which we had by us, and Eggs or Silver Gobbets came out, White as Salt, these we tinged, with our Saffron: of them we fold publickly two hundred times, with which we have been made Rich and our Treasures are multiplied.

XI. And whosoever you shall imbue or fill with the Powers thereof, should they be hurt with the Poyson of Vipers, or the Malignity of Brass or Verdigrise, they shall be in no danger; for that it quickens and revives the Dead, and kills the Living: It destroys and restores again; it casts down that which is elevated and lift up, and elevates that which is abjected and cast down, and gives you a dominion over the Heavens of the Earth.

XII. Now you must note, that there are two Stones of
the Wise, found in the Shores of the Rivers, in the
Arms of the Mountains, in the Bowels of the Floods,
and in the back parts of the Kings House, which by
instruction and prudent management may be brought
forth, Male and Female.

XIII. By these being conjoyned and made complex (or
perfectly united into one consimiliar substance) you
will be made wiser (you will see the reason of the
Operation, and the end of the Work) Blessed God, how
great and how wonderful a thing is this.

XIV. A certain Philosopher dreamed, that the Kings Messenger came to a certain Podagrick, and the Podagrick
desired that he might go with him; to whom the
Messenger answered, since thou hast the Gout, how
canst thou go with me, for thou canst not walk.

XV. To whom the Podagrick answered, Thou knowest that
in the Root of this Mountain, there is a certain Tabernacle, bearing me then thither, leave there the
burthen, so shall I be presently freed or delivered
from my Gout.

XVI. Then said he to me, thou art not able to touch the foot thereof: but going back, he took him up and placed him in the Tabernacle, the foot of which, the Messenger said, he was not able to touch: And waking from his Dream, he saw nothing. Behold the Similitude.

XVII. Another also saw in a Dream, wherein it was said, if any one truly should sit down by the way, and should ask you, whether you would think fit to do this thing, would you do it? He answered, I know not; the other said, thar he should lie or generate with his Mother in the middle of the Earth; then awaking, he saw nothing. Consider well this similitude.

CHAPTER XXI

THE RETAINING OPERATIONS, AND CONCLUSION OF THIS WHOLE WORK.

I. But leading you to the knowledge of Phylosophy, and exposing the Demonstrations thereof, in a Philosophick manner, we should make it the dirision and mockery of Women, and the play of Children.

II. Take also the fresh Bark or Rind, in the same moment; in which you shall after another manner, extract the matter or thing it self, in the place where it is generated; and put it in to a Cucurbit, and sublime it.

III. And that which is or shall be sublimed, separate it, for it is the Vinegar of the Philosophers, and their Sapience (i.e. their Salt.)

IV. Then take this Vinegar, and melt or pour it forth upon another Cortex, Bark, or Superficies of the Sea, and put it into a Glass Vesica, in which put so much of your Vinegar, as may over top it the heighth of Four Inches; this bury in warm Horse-dung, for Thirty One (or Forty) Daies.

V. This time being past, take the Vessel forth, and you will find it now dissolved, and turned into a black and stinking Water; more black and stinking than any thing in the whole World.

VI. Then take this very thing it self, and very gently elevate it in its tabernacle, till all the moisture is consumed, so as no more will ascend, this sublimed Matter keep carefully for your use.

VII. Then take the Foeces, which remain in the bottom of the Cucurbit, and keep them, for they are the crown (and rejoycing) of the Heart. Dry then the same and grind them, and add there to fresh or new Cortex of the Sea, that is say, Mercury, and grind them together, drying them in a warm Sun.

VIII. And the Waters from the same first sublimed, sink down to the bottom, which diligently grind and dry, and put them in the Crucible or Test of Ethel, and sublime: and the Matter being sublimed purely White, as fine Salt, keep it safely, for it is the Auripigment, and Sulphur and Magnesia of the Philosophers.

IX. Understand now, and see that you govern your Work with Wisdom and Prudence, and make not too much haste.

X. Then take the Cucurbit, put half way into lute, and put into the same, your dissolved black Water, which you have sublimed; that is to say, nine parts, and of this whitned Auripigmentum, which you sublimed from the Ethel two parts.

XI. I say, that this opened or decocted Auripigment, is immediately dissolved in the Water, and made like to Water; that nothing can be seen by mankind, of a more intense, fixt, and perfect Whiteness, nor any thing more beautiful to the Eye, which the Philosophers call their Sal Virginis, or Virgin Salt.

XII. Put this into a little Vessel, called a Cucurbit, close well the Joynts, which put upon a gentle Fire, making it, as it were, but with two Coals at first, and then adding two others: and look into it, to see how the Water ascends and descends.

XIII. When you see the Vapour is consumed, and nothing more will ascend, of that which is elevated nor

179

descend, know that the Matter it self is now coagulated: make therefore a more intense and vehement Fire, for the space of three hours of the day.

XIV. Then lastly, take away the Fire, or let it go out, and the next day (all things being cold) open the mouth of your Cucurbit, and take forth the Matter, which is of a substance, white, sincere, and melted or dissolved.

XV. This is your Substance sought after; and now you have come through to the end of your Work; manage it according to your Reason and Prudence, for (God assisting) you may make of it what you please.

FINIS

CYLIANI

hermes

unveiled

RAMS

TRANSLATOR'S NOTE

The man known as Cyliani, of whom little is known personally, wrote the present volume in 1831 and had it published the Year after. Its main interest lies in the fact that he influenced a school of French Alchemysts who based their work On his findings, the first of these was G. F. Tiffereau, who brought a piece of gold from Mexico, which he claimed to Have manufactured by the art. Tiffereau spent the years from 1847 to 1891 trying to persuade French scientists to take his work seriously, but only succeeded in stimulating The Alchemysts such as Jollivet-Castelot and others. He complained that the Sun in France was not as suitable to the work as that of Mexico.

This is the only translation of the book into English of which only one copy, the present, has been made.

HERMES UNVEILED

BY

CYLIANI

Translated by: *Ivan Cordet*

PREFACE

Heaven having granted me success in making the Philosopher's Stone, after having spent thirty-seven years in seeking it, stayed awake at least fifteen hundred nights without sleep, suffering innumerable miseries and irreparable losses. I have decided to offer to youth, the hope of the future, the heartrending picture of my life. This may serve both as a lesson and at the same time help the young to turn away from an art that at the first sight may offer the most delightful white and red roses that, however, are surrounded by thorns and of which the path that leads to the place where one can pick them is full of pitfalls.

The universal medicine being a far greater blessing than the gift of riches to know it naturally attracts studious men who believe themselves to be happier than the multitudes. This reason has influenced me to transmit to posterity the processes to be undergone in the greatest details, without leaving out anything at all, in order to let it be known and to prevent the ruin of honest people and to render a service to suffering humanity.

The reader who understands my processes will only have to seek the Matter, the Fire, and the Labors of Hercules. Duty has forced all philosophers to make a mystery of this. I have sworn to Almighty God to carry this secret into the grave and will not be perjured, were I even to be stoned, as I prefer to displease men rather than the eternal.

I have honestly described all the processes that have to be undertaken, on this you can count. I have hope from the depths of my heart that what I have dreamed will reveal to virtuous man, if so called by God, how to rejoice in such blessing, helping them to avoid the numberless traps, even to the possibility of losing his very life.

The greatest care is needed to pass the Labors of Hercules. Once they are passed the rest is a very agreeable work that does not cost one cent in money.

Be very careful that it does not happen to you, as it did to me, to be wounded. As a result of these varied labors I find myself with the most essential organ of life affected, which means that I shall be denied, considering the seriousness of the illness, enjoying a long life, the virture of the medicine not being surgical, but only medicinal.

I would advise those who insist on treading the path that leads towards the Philosopher's Stone, only to begin their journey after having followed several courses in chemistry so that they will know how to manipulate. Whatever many authors may say, if I had not had the knowledge of chemistry that I do possess, I would never have reached the goal.

I must add that the matter proper to the work is that which served to form the body of primitive man. It is to be found everywhere and in varied forms. Its origin is both divine and terrestial, equally so the fire of the stone.

The Universal Medicine is a magnetic salt that serves as a covering for a strange energy, universal life. As soon as this salt reaches the stomach it penetrates the whole body to its ultimate limits, regenerates all its parts, brings about a natural crisis followed by abundant sweatings, purifies the blood as well as the body, strengthens this matter instead of weakening it, through dissolving and chasing away by perspiration all morbid matter which would impede the play of life and its currents. This salt, due to its cold quality, also makes disappear every kind of inflammation while the mysterious energy of this same salt spreads itself throughout the principal organs of life, settles there while vivifying them. This is the effect of the Universal Medicine which medically cures all infirmities that may affect man throughout the course of his life. It allows him to exist for several centuries in good health, unless, due to his constitution, God has ordered otherwise. This effect is quite contrary to the opinion we obtain from doctors who insist that one remedy only cannot cure all illnesses. But if they were cognizant of the Universal Medicine they would see that the power of this salt is similar to that of a magnet that attracts not in this case iron, but the energy of universal life for which it acts as a covering. In administering it, they would be forced to recognize its celestial power, and they would fall onto their knees before this beautiful magnetic salt endowed with such miraculous and supernatural power and at the same time would declare from the depths of their hearts that no illness can resist its action. Of this I have been able to convince myself by giving back life to many sick people already abandoned by doctors.

To understand what I have just said regarding this strange power within the medicine more clearly, one should observe the effects of champagne in our stomachs. As soon as it enters the stomach, its liquid penetrates the walls and strengthens them. At the same time, its unknown energy, due to the presence of carbonic acid, is freed and rises to the brain, chases away our sorrows and makes us feel gay unless, of course, too large a quantity of this gas starts pressing on the brain and makes us stagger or fall.

We must realize that the body of man is material; nevertheless, it contains a power and energy which is foreign to it. This energy is life.

At this point, I must warn you never to forget that only two matters of the same origin are needed: One volatile, the other fixed. That there are two ways, the dry and the humid. I personally follow the latter by preference and by duty, though the former is familiar to me. It is done with only one matter.

The azoth unites easily with sulphur, fire with fire, and the double mercury or rebis in powder or

188

sale or oil forms the true potable gold or the Universal Medicine in white or red. Finally the seed of gold lies within the gold itself.

Few combustibles are necessary; even less receptacles. The work costs very little to undertake and can be performed in any place, but it is convenient to begin it with that of nature in order to finish it well. I thought in this book to retain the most important passages from several works written by those philosophers who have best grasped the secret mercury, that is to say, Hermes, Arnold of Villanova and the anonymous author printed in Leipzich in 1732 and others, so as to transmit in a primordial manner this most precious and divine art relating to health.

Seek to know the Vinegar of the Mountains because without it you can do nothing. Its knowledge will give you that of the fairy of the soul so called by Arnold of Villanova in his Little Rosary.

Also, consider deeply that the fire of our hearths, that of the furnaces, or of a lamp is the tyrant of destruction and that nature only uses the common fire in order to destroy. For example, the fire of lightning or that of the volcanos.

Remember that, after their preparation, the two metallic natures cannot be reunited, but only in the state of dissolved germs, as is told by Arnold of Villanova.

Having well understood the practice and the workings that I am going to give you, you can apply

it and that only the elect, whom God would please to initiate, would be allowed to possess this supernatural knowledge. I was, by nature, good and a believer. I was simple and did not know the twistings of the human heart. I believed in the sincerity of these books. I could hardly wait to become my own master so that I could devote myself entirely to this type of study. From my point of view, life only had charm so long as one was possessed of health and could ~make others happy without their being able to speak about you. The knowledge of the Philosopher's Stone would achieve this goal. It became the main consideration of my waking moments and of my leisure. My ambition was also to obtain the certainty that the soul existed and was immortal. Such was the knowledge I wished to acquire, even if it meant losing my very life.

The French Revolution had just burst forth. My knowledge seemed, in the eyes of my co-citizens, more useful in an administrative post than in the army. I was honored with several posts. During one of my tours, I saw when entering a small town a beautiful young lady whose appearance of goodness, sweet smile and air of general decency charmed my soul and inflamed my heart. At that very moment I resolved to make her my wife. After having fulfilled my mission I busied myself in finding a pretext to speak to her. Love was not lacking and not many days passed when the moment came that I received permission to visit her.

Finally she accepted me and I promised to make her the happiest woman in the World. Alas! I was a long way from believing that I would be the cause of her experiencing a series of unprecedented miseries when considering that she did everything to make me happy.

A few months after my marriage, I made the acquaintance of a talented man whose wife was a famous artist. They were both interested in alchemy and trusted me with a little manuscript that had been found behind a closet. They were deeply impressed by this manuscript. It was written in a style that inspired a great deal of confidence. Everything was to be found there, with the exception of the name of the Matter, of the Labors of Hercules and the knowledge of the Fire. I had felt I had reason to believe myself the happiest man on Earth. In the hotheadedness of youth, I became to conceive immense projects. I put myself to work which made me neglect my profession and personal needs. I then decided to resign from my post in order to be able to concentrate on Hermetic philosophy and within a few years I had used up all the money that my father and mother had given me upon my marriage and also a part of my wife's dowry had been flown up the chimney.

My unlimited love and friendship for the companion of my youth and her tender feelings for me provided us with a large family that increased my expenses as my personal fortune diminished. I saw my wife bravely putting up with her situation and the desire to render her happy greatly increased my firm resolve to achieve the goal upon which I had decided. Twenty-one years passed thus in a state of constant hardship. I fell into the deepest misery. My many friends now turned their backs upon me. Everyone lost interest in me, though some tried to discover why I had fallen into this sad state, especially considering my exemplary conduct. It came to light that my taste for alchemy deprived me of even the basic necessities. I became an object for public laughter. I was treated as a madman. I was booed. My family rejected me and on several occasions I found myself wandering over the country, obliged to interrupt my work, having sold even the best of my clothes in order to pay the wages of a servant who helped me pass my nights. My wife, burdened with many children, was obliged to take refuge with her parents, yet never ceasing to be a model of all the virtures. And, looking into the depths of my heart, I found nothing with which to reproach myself beyond my taste for a work that had ruined me and placed my family in such a painful situation.

I found it necessary to forget my work and to sell my talents, but the poor situation in which I stood made me unwanted. Hardly had I organized some advantageous deal that my subordinates or some people who furnished funds seized them, then seeking to discredit me so that I could find no help. After about ten years has passed thus and having made use of a part of every night reading practically all the works published on the Philospher's Stone, my head beginning to be bowed by the weight of years, I began to feel the irrestible drive that reminds man of his first loves. In good faith, I suddenly believed myself better instructed than others and capable of overcoming all the obstacles that had stopped me heretofore. I had recourse to rich people of similar tastes. I was received in a very friendly manner. I passed happy days during the early days with these new acquaintances. Friendship was lavished upon me and I could now, to some extent, come to the help of my family. However, as soon as they thought that they had obtained all that I knew from me, I was abandoned with futile excuses. It even happened that some went to the point of giving me a strong dose of corrosive sublimate to destroy me and steal my writings. I learned to know the human heart at great cost. I had to keep wide awake all the time. But the fire which manifested in my stomach and the taste I recognized immediately, made me run to an antidote. This resulted in a year of discomfort and being almost deprived of the only pleasure that I knew on earth. In order to avoid importunating the reader and being accused of wordiness, I shall avoid a recital of the more petty human passions that I observed and the inconceivable differences that exist between the pleasant man one meets decorating the evenings in our drawing rooms and the same man controlled by the lure of riches and by his own vile cupidity. They are truly two different people.

I do not wish to pursue the recital of what happened to me. A description of the misfortunes that befell me would require a large volume. I once more fell into hardship. It was such that my numerous family, made up of charming well educated children, virtuous beyond expression by their decency and their talents were so unhappy because of the misfortunes of their father that they caught what in others would have been but slight diseases and, after a couple of weeks, these diseases became fatal for them. In a short while I lost my children.

Oh irreparable loss! How sad and heartbreaking it is for the paternal heart to be able to weep but tears and to feel only superfluous regrets! May the Eternal one day allow me once more to see you again and then the memory of my misfortunes will be erased for me. Despite the broken-down condition in which I found myself, I decided to re-gather my strength in order to make one last effort. I went to see a rich person who was endowed with a great soul and equally great culture. For several years I was treated by her in a most generous manner. Far more generous than the others whom I had met in the past. I managed in the end to produce something encouraging, but it was not yet the true work.

One day as I had been walking in the country and was now sitting at the foot of a large oak tree, I took pleasure in bringing to my mind all the circumstances of my life and trying to judge whether I had some merit whatsoever or if I had deserved the huge weight of hardships that had brought me low. I remembered the discoveries I had made for commerce and the benefits that French industry had obtained from them. It was painful to see that strangers had profited and my name had been forgotten. I remembered the people who had had the skill to take advantage of all of their discoveries after given them a more fashionable form. They were covered with honors well situated while I was wandering and rejected. I asked myself if I had ever intentionally robbed one cent from any one of my fellow men and my conscience answered in the negative. Did I for one moment cease being a good son, good husband, good father, a good friend to those who deserved it? My heart answered me again: No, your unhappiness stems only from not having achieved your goal.

I brought to mind that it had been a cruel thing that so many times I had been so ill paid by my fellow men, even by my friends. I felt broken down by the pain of these memories. My strength and energy left me and I put my head on my hands and a very torrent of tears poured forth, as at the same time I called the Eternal to my help. That day it was very hot, so that I fell asleep and then dreamed a dream that I shall never forget.

I thought that I heard split the true at whose feet I was sitting. The sound made me turn my head. I saw a nymph, a very epitome of beauty emerging from this tree. Her clothes were so diaphanous that they seemed transparent. She said to me: "From the heart of this holy tree I have heard the repetition of your misfortunes. There is no doubt but they were great, but such is the fate toward which ambition leads a youth who seeks and who believes himself able to face all dangers in order to satisfy his desires. I shall add no comment in order not to aggravate your misfortunes which I can ease. My essence is celestial. You can even consider me as a ray from the pole star. My power is such that I animate everything. I am the astral spirit. I give life to everything that breathes and vegetates. I know everything. Speak what can I do for you?"

"O celestial nymph," I said, "you can reanimate in me a heart beaten down by hardship by granting me only some slight idea regarding the organization of the universe, regarding the immortality of the soul and give me somehow the means that I may achieve to a knowledge of the Philospher's Stone and the Universal Medicine. I have become an object of public laughter. My back is bent under the huge weight of miseries. For pity's sake deign give me the means to rehabilitate myself in my own eyes."

"I am truly touched by your sorrowful existence," she answered me. "Listen. Gather all your faculties and engrave into your memory what I am going to tell you. At the same time, take a part of my comparisons as figurative, in order that I may be able to make myself understandable to your intelligence."

"Place before your mind an immense space, almost without limits, within which floats this system of worlds composed of suns, of fixed stars, of nebulae, of comets, of planets and satellites, all swimming within the breast of eternity, all within the light of a divine sun whose rays are without limits and you will have some slight notion of the totality of the universe, as well as both the finite and the infinite world."

"This system of worlds and of the Eternal or the sun of divine light are all of the same origin. They have had neither beginning nor will they have an end. The slight changes that certain globes experience changes nothing in the order of the universe."

"The will of the Eternal or of the Creative Spirit can at will hurl a nebula into space. This going off at a tangent while flying through space is subject to the law of attraction of any sun to which it has drawn near and finishes by describing a very elongated eclipse whose two nuclei are determined by the action of the two suns. Then it is transformed into a comet but at the end of a lapse of centuries it finishes by giving way to the attraction of the stronger of the two suns. Its course becomes regular and finishes by making part of its own solar system by

turning in circles around it. Then after a certain
number of centuries its two luminous points become one
point only, one luminous point that becomes the
central fire of this globe, which itself becomes at a
very, very distant time a planet habitable as soon as
it had taken on a certain metaliferous consistency.
Then it brings to birth on its surface the elements
necessary to the life of the animals proper to its
nature, such for example as water and atmosphere and
vegetables."

"The planets can, by a powerful expansion of the
central fire, tear themselves into different parts,
each of which spreading out into space, becomes also a
satellite that is attached to the active atmosphere of
some other planet."

"A comet which, in the first place, was a nebula
can, by its effect when drawing too near to a planet,
raise its waters and give rise to a flood by a
lowering or raising its axis. This changes the bed of
the seas and brings to light that which was covered by
the waters and hidden for centuries under them such as
inhabited countries which recovered from the ooze
of the sea bottom raising at the same time the debris
of animals and of vegetables packed tightly one upon
the other."

"Another planet, passing through the tail of a
comet, may have its atmosphere so burnt up that not
only all animal and vegetable life is destroyed, but
the planet itself is transformed into a vast tomb.

Finally, a comet, by its excessive action when drawing near to a planet, can cause such a disturbance on it that animal and vegetable life may be greatly modified, even destroyed. These are the only alterations that happen to the globes, but, for all that, nothing in the world is lost. Were these globes to be reduced to atoms, these last, by the law of magnetic attraction, would end up by bringing forth a new planet."

"The various species of animals that seemed to have existed in the world at distant successive epochs, are the facts of the creation brought about by the Creative Spirit. But, all the beings who come from these appear at epochs more or less distant the one from the other. They seem to be the results of great catastrophes that happened to the earth. The human species itself only seem to date about sixty centuries."

"The suns, the comets and the various planets are as much beings of a specific nature that find themselves definitely ruled by a spirit. The universal hierarchy is infinite. The Eternal belongs to a category far above these spirits. These latter are, as it were, its ministers and the planets, the globes, its subjects that have to submit to the directives of the same ministers."

"Everything that exists in the universe that is material or physical or purely mineral, even the gases, note this avowal.

"Man is triune, His body or his shape is animated by a soul. This is the junction of diverse forces by the help of which the spirit imprints its shape or matter. The soul is directed by the Celestial Spirit which is an emanation of the divine action and, consequently, imperishable.

"Only man's form perishes. The spirit for which the soul serves as a tie or an envelope, separates itself and man's form, deprived of this vital celestial spirit, is delivered to the reaction of its constituent principles. Spirit and soul live then spiritually in seeking the centers that are suitable to them and, after the lapse of a certain time, man or the being or the spirit or the spiritual life which goes on always perfecting itself, separates itself from his soul or from his glorious envelope to enter into its universality. This means that man dies twice. That is to say, twice changes his form. But man or the spirit lives eternally. From what I have told you, you can now no longer doubt the immortality of the soul.

"This is all that I am allowed to teach you now to satisfy your desires. "Now do you want to know how the Universal Medicine acts on the animal economy? Consider as I have just told you how only man's form is mortal and you will see that only the solid aspect perishes. As these latter are all mineral, all can be regenerated by the principle or mineralizing spirit which, by its various modifications, forms the various products we know. They are all returned to their primordial state by the action of this same principle and by its mysterious power which re-establishes the balance and allows the spirit freely to enter and to leave through our proper form just as water passes through a sponge. Because the upsets of our body only come, excepting mechanical disturbances, because the currents of life cannot circulate freely. But the virtue of the Universal Medicine is purely medicinal and not surgical. It cannot replace an amputated limb or one that is entirely destroyed, which means that the person who takes it in good time, generally at the two equinoxes, can live without infirmities for several centuries unless nature has predestined a short duration of his life due to his physical make—up which never ceases to combat the efforts of life.

"Let us now come to this matter of all your miseries and, if I dare say so, of your fixed point. Your obstinacy was required in order to make you worthy of such good fortune. Listen attentively and never forget your hardships so that throughout your life you will always remember those who are unfortunate. Follow me and fear nothing."

I then saw a cloud that seemed to emanate from the center of the earth. We were wrapped in this cloud and it carried us into the air. We wandered along the seashore on which I noticed small humps. Night came. The sky was covered with stars. We followed the Milky Way, directing ourselves toward the pole star. An extreme cold caught hold of me and brought on a deep sleep. Warmed afterwards, due to the rays of the sun that began to appear on the horizon, I was very astounded on waking up to find myself on the earth and here to notice a temple. The nymph took me by the hand and led me to its entrance.

"You have now arrived," she told me, "at the spot where you must resolve the following problem. As you have been a good mathematician, reflect well because you can do nothing without solving it. From one by one which is only one are made three, from the three, two, and from the two, one.

"You have told me that you have been educated in chemistry. Look and see what means your knowledge can offer simply to open the lock of the door of this temple in order that you may penetrate into its very sanctuary.

"A victory won without danger," she said, "is a triumph without glory. Before leaving I must once more point out to you that you can only battle the dragon that defends interiorly the entry of this temple, with this spear which you must make red hot by the aid of a comon fire in order to pierce the body of the monster which you must fight and to penetrate right to its very heart. This dragon has been well described by the ancients and they have spoken of it on many occasions."

"Think of the dew of May. It becomes indispensible as vehicle and principle of all things."

I looked at her; the nymph began to smile. "At last you will begin the labors of Hercules. Collect all your energies and be of firm will. Adieu." The nymph took my hand and squeezed it. "Do you love life?" she said to me.

I answered her, "In your presence I cherish it more than ever." "Try not to lose it as a result of imprudence. Waiting for the results of the battle I shall be watching by you and in case that anything happens, I shall come and help you. Adieu." She disappeared.

I was so sad to have lost this nymph who had now become so dear to me. Finally I made up my mind and made ready for the battle. Having collected branches of dry wood that were scattered on the spot where I found myself, I lit them with the help of a lens that I found to be on me and heated my spear almost white hot. While doing this, I sought the means which would best destroy the lock on the door of the temple. I noticed that the nymph had slipped into my pocket without my noticing it a corked flask full of the substance that was necessary to me.

Determined to win or to perish, I furiously seized my spear with one hand and the substance in the other and put a sufficient quantity of this latter on the lock. In a little while the lock disappeared entirely and the two leaves of the door to the temple opened with a loud noise. My eyes fell on a fierce dragon who was endowed with an enormous three pointed tongue with which he sought to throw his fatal breath on me. I hurled myself towards him crying out: "When one has lost everything, when one has no more hope, life becomes a disgrace and death a duty."

He opened his huge jaws to devour me, at which I hurled my spear with all my strength so that it pierced through his throat, deep into his entrails. I tore out his heart and, so that he could not reach me, at the same time, I made crude efforts with the help of my spear to turn away the direction of his head. The monster curled up on himself several times, vomited waves of blood and ceased to exist.

Following this I walked to the heart of the temple and there heard a celestial voice speak, saying to me: "Audacious one, do you dare profane this temple in order to satisfy your vile cupidity, or do you come here in order to seek the means to help suffering humanity?"

I answered, "I come free from all ambition to pray on my knees for you to grant me only the return of the fortune that I have sacrificed in order to know the Philosopher's Stone. Also the means so that I can secretly help virtuous human beings. I swear to you, and I swear it before the Eternal, that if you deign to accord me such a boon, I shall never reveal the Labors of Hercules nor the Matter nor the Fire, except by means of a language that could only be understood by those whom God would wish to trust with such a secret and if I am perjured, may I be punished in the proper manner."

I then saw two superb crystal vases each resting on a pedestal made of the most beautiful marble of Earrara.

One of these vases had the shape of an urn surmounted by a gold crown with four fleurons. On it had been engraved the following words: Matter containing the two metallic natures.

The other crystal vase was sealed with a glass stopper. It was very thick and on it was engraved similarly that which now follows: Astral spirit or ardent spirit which is a projection from the pole star.

This vase was adorned with a silver crown decorated with nine brilliant stars.

As I finished reading, I noticed with great joy my sweet nymph who said to me in showing me the large flask, "Do you see my mirror? Nothing," she said to me, "can put itself in opposition now from your rewarding yourself for the struggle that you have sustained with so much courage while taking at will the substances that are contained in these two sacred vases which are of the same celestial origin. I noticed the discomfort that your victory is causing you. This could become dangerous to you if you stay too long in this place. Hasten to take your reward and leave this temple with all possible speed. I shall prepare everything for our departure." She left me alone.

My strength and courage began to wane. I felt I must obey the orders of the nymph. I noticed at the sides of the two sacred vases various empty flasks, very neatly made in crystal and corked with glass stoppers. I took two of them, opened hastily the first vase in the form of an urn that contained the androgyous matter and the two metallic natures and filled my container with it. After having sealed the crystal urn, I opened the second and larger vase and tremblingly poured into my second flask some of the substance that it contained. I had no funnel. Time lacked, my strength was failing, I quickly closed the large vase and my own with its crystal cork and hastened from the temple. In passing near to the monster I had conquered, I saw that nothing was left of him except his mortal form which was of no value whatsoever.

No sooner had I come out into the fresh air than I thought that I would faint. Fearing to break my two vases through falling, I lay down on the earth with great caution having been careful to place my two flasks beside me. For a few moments I found it hard to breathe. My cherished nymph came to me, smiling. She congratulated me on my courage and on the victory I had just won. She said to me, "Be sure, unfortunate Cyliani, that it is not good to expose yourself frequently to such a battle. What am I seeing?" she said to me, "A school!"

Her words struck me. I said to her, "Explain yourself."

"One of your vases," she said, "seems to hold more androgynious matter than you will need, but you have not taken enough of the Astral Spirit. You need infinitely more. And, as Arnold of Villanova said, it is necessary to have an abundance of water, of distilled spirit, but your fault is excusable. It is the result of a valid fear. At any rate, you have enough to teach you how to make the Stone and how to achieve all your desires. Let us now hasten to return to our point of departure. You do not seem to be thinking any more of the companion of your youth, nor of the anxiety into which your absence has plunged her. Let us go. Here your life will be in danger."

I saw a new cloud come out of the center of the earth. We soon found ourselves wrapped in it and lifted into the air. We made good speed. Night overcame us. The sky was without a cloud and well covered with stars. We again followed the Milky Way, but in a contrary direction. then began to feel extremely cold. Our direction was also on the side of the place which saw my birth. But, in leaving a cold region and in passing into a hot region, I felt a deep sleep overcome me and was astonished when I woke up at the aurora at sunset, to find myself at the foot of the large oak from which we had left.

I called to my amiable nymph and she said to me, laughing, "What do you want more?"

Then I replied, "Tell me what it is necessary for me to do in order to finish my work."

"Now that you have passed the Labors of Hercules and that you possess the Matters, it is nothing more than a work that could be done by a woman or careful and attentive child. Listen closely."

"Observe the working of nature. In the center of the earth she has formed the metals, but something else is necessary -- their quintessence. Observe from whence she draws the quintessence of things. It is only on the surface of the earth, in the kingdoms that live or vegetate. Just follow nature, step by step. Consider also how she functions in the vegetable kingdom, because it is not a mineral that you are going to make. Observe how she dampens with dew or with rain the seed that has been entrusted to the earth. How she dries it with the aid of celestial fire and reiterating thus until the embryo is formed, developed, buds, flowers, grows and as it reaches its multiplicative virtue. Finally, it matures as fruit. It is very simple. Dissolve and coagulate. That is all. And be very careful not to use any other fire than that of heaven."

Finally the nymph was so kind as to designate everything that I still had to do, as I shall tell in the greatest detail. I threw myself at her feet to thank her from such an unparalleled blessing while offering my humble thanks to the Eternal for having allowed me to overcome so many dangers. Then she said goodbye to me, adding, "Do not forget me."

She disappeared. Her flight made me experience such great pain that I woke up.

A little later I started my work again, now aided by the Labors of Hercules. I was able to produce the matter than contains the two metallic natures, as well as the Astral Spirit. This with the help of my last resources and not those of other times. This made me free to handle my success as I would towards those who deserved it in my eyes, without hurting my sense of delicacy or my comfort, nor bringing out unnecessary gratitude.

FIRST PROCEDURE
PREPARING OF THE AZOTH
OR THE PHILOSOPHER'S MERCURY

I first took some of the matter that contained the two metallic natures. I began by imbuing it, little by little, with Astral Spirit in order to awaken the two interior fires, which were as it were put out, by lightly drying and stirring the whole with a circular movement by the heat of the sun. Then, reiterating thus and frequently damping more and more, drying and stirring, to the point that the matter had taken on the aspect of a slightly thick porriage.

Then I poured on top of it a new quantity of Astral Spirit, just enough to cover the matter and left the whole in this condition for five days, at the end of which I adroitly decanted the liquid or the solution, which I then kept in a cold spot. Then the matter that was left over I dried in the heat of the sun, in a glass vase which was approximately three fingers in height. I imbued, I stirred, I re-dried and dissolved as I had done before and reiterated thus until I had dissolved all that it was possible to dissolve, having been careful to pour each solution into the same well-sealed vase which I then put for ten days in the coldest place that I could find.

When ten days had passed, I put the whole solution to ferment in a hold-fast for forty days at the end of which, as a result of the internal heat, it precipitated, as a result of the internal heat of the fermentation, a black matter.

It is then that I distilled, to the best of my ability, the precious liquid that floated on top of the matter containing its interior fire and placed it in a vase of white glass, well stoppered, in a place that was both humid and cold.

I took the black matter and dried it in the heat of the sun, as I have already said: Reiterating the imbuings with Astral Spirit, but stopping immediately that I noticed that the matter was beginning to dry up and thus leaving it to dry out by itself. This was done as many times as was necessary so that the matter could become as a shining black pitch. Then the petrification was total and I stopped using the exterior fire so as not to damage the matter through burning the tender soul of the black earth. By this means, the matter came to imitate horse manure. It was necessary according to what we are told by the philosophers to let the interior heat of the matter act upon itself.

It is now necessary to begin again, using the exterior fire, in order to coagulate the matter and its spirit. After having allowed it to dry itself, one imbues it little by little with its distilled reserved liquid which contains its own fire. The matter crushed should be imbued and dried by a light solar heat until it has absorbed all its water. By this means, the water is entirely transformed into earth and this

214

latter, through a process of dessication, changes itself into a white powder that we also call air. This falls like a cinder containing the salt or the Mercury of the Philosophers.

During this first procedure one sees that the solution or water is changed into earth and this by a form of subtilization or sublimiation transforms itself into air by means of the art and here is where one stops the first work.

By means of the new Astral Spirit, one dissolves this ash little by little leaving, after the solution and the decanting, a black earth that contains the fixed sulphur. But in reiterating the operation on this last solution, exactly as we have described, an earth is obtained that is even whiter than the first time, which is called the first eagle. One reiterates this process seven or nine times. By this means one obtains the Universal Menstruum, or the Mercury of the Philosophers or the Azoth with the aid of which one extracts the active and personal energy of each body.

It is here good to observe that, before passing from the first eagle to the second as well as to the following, it is important to reiterate the preceding operation on the ash that has been left, if the salt is not sufficiently elevated by the central fire of the matter to the philosophical sublimation, so that nothing is left after the operation other than a black earth from which its mercury has been removed.

Be very careful here to note that following the swelling of the matter during the process of fermentation that follows the solution that at the upper part of the matter a sort of skin forms on which are to be found a mass of little bubbles which contain the spirit. It is at this point, that the fire must be led in a prudent manner as the spirit tends to take on an oily form and passes to a certain degree of ciccisity.

As soon as the matter is dissolved it swells up, begins to ferment and gives out a slight noise which proves that she holds within herself a vital germ which frees itself in the form of bubbles.

In order properly to undertake the process I have just described, it is necessary to observe the weights, the control of the fire and the size of the vase. The weight must consist in the quantity of Astral Spirit necessary for the solution of the matter. The handling of the external fire must be directed so that the bubbles that contain the spirit do not evaporate in too large a quantity and do not burn the flowers or the sulphur by continuing the external fire so do not push the drying of the matter too long after its fermentation and its putrefication in order not to see the red before the black. Finally the size of the vase must be based on the quantity of matter, so that the vase only contains a quarter of its capacity. Do you understand me?

Also do not forget that the mysterious solution

of the matter or the magical marriage between Venus and Mars takes place in the temple of which I spoke to you previously on a fine night with a calm cloudless sky, the sun being in the sign of Gemini, the moon being in its first quarter at its full with the help of the magnet which attracts the Astral Spirit from the heavens. It is seven times rectified to the point that it can calcine gold.

Now the first operation being ended, one has the Azoth or the White Mercury or the cell or the Secret Fire of the philosophers. From this point on certain sages dissolve it the least quantity of Astral Spirit necessary to produce a thick solution. Having the solution, they place it in a cold place in order to obtain the three layers of salt.

The first salt has the aspect of wool. The second of a niter with very small points and the third is a fixed, alcaline salt.

Some philosophers use them separately; others bring them together as is pointed out in A. de Villanova in his *Petit Rosaire* written in 1306, in the article on the "Two Leads" and dissolves them in four times their weight of Astral Spirit in order to achieve all their operations.

The first salt is the veritable Mercury of the Philosophers. It is the key that opens all the metals by the aid of which one extracts their tinctures. It dissolves everything radically. It fixes and ripens all equally while fixing the bodies by its cold and congealing nature. Briefly it is a very active universal essence. It is the vase in which all philosophical processes are undertaken. One thus sees that the Mercury of the Wise is a salt which is named dry water which does not wet the hands, but to make use of it, it must be dissolved in the Astral .Spirit, as we have already said. One uses ten parts of mercury against one of gold.

The second salt serves to separate the pure from the impure and the third salt serves continually to increase our mercury.

SECOND PROCEDURE
PREPARATION OF THE SULPHUR

The tincture extracted from common gold is obtained through the preparation of its sulphur which is the result of its philosophical calcination that causes it to lose its metallic nature by changing it into a pure earth. This calcination which cannot-take place by means of the common fire but only by using the Secret Fire existing within the Mercury of the Wise, because of its dual nature. And, it is by virtue of this Celestial Fire-supported by trituation which penetrates into the center of common gold that the dual central fire of gold, mercurial and sulphurous, which is found therein as if dead and imprisoned is now discovered to be untied and animated. The same Celestial Fire, after having extracted the tincture of the gold, fixes it by means of its cold and coagulating quality. It then becomes perfect, capable of multiplying itself in both quality as well as quantity. This earth having reached a state of stability takes on the color of peach blossom, which gives the tincture or the fire which is then the vital and vegetative gold of the wise. This takes place by means of the regeneration of the gold through our mercury.

It is now necessary to dissolve the common gold into its spermatic matter by means of the water of our mercury of our Azoth.

In order to reach this point the gold must be reduced to a very pure calx or oxide of a browny-red color. After having washed it over a low fire several times with well distilled rain water, it is necessary to dry it lightly by the heat of the sun. Then we calcify it by means of our Secret Fire. It is upon this occasion that the philosophers say: "The chemists burn with fire and we with water."

After having soaked and lightly crushed the oxide of well calcined gold and having it drink its own weight in salt, or dry earth which does not wet the hands, and having well incorporated them together from now they are wetted by successively increasing the dampings to the point that the whole resembles a slightly thick pap. Then place on top of this a certain quantity of water of mercury, proportionate to the matter, to the point that it just covers this last. The whole is now left for five days within the gentle heat of the water bath of the wise. At the end of this time the solution is to be decanted into a well stoppered vase and which will be kept in a damp and cold place.

The matter that has not been dissolved is then to be taken and left to dry in a heat similar to that of the sun. Once it is dry enough, the frequent dampings and triturations are started once again in the same manner as previously indicated, in order to obtain a new solution which will be reunited with the first.

Repeat in this manner until you have dissolved everything that can be and nothing is left except the dead earth, which is of no value. The solution being finished and collected in the well-sealed vase of which we have already spoken, its color should be the same as that of a lapis lazuli. This vase is to be placed in the coldest possible place for about ten days. Then the matter will be allowed to ferment as we have already said in the first procedure and, by the proper internal fire of this fermentation, a black matter will be precipitated. This matter is to be distilled skillfully and without fire by putting the liquid separated by the distillation which floated on top of the black earth into a well-sealed vase in a cold place.

Take the black earth, separated by the distillation of its liquid, and allow it to dry by itself. From now on it is permeated by the external fire: that is to say, by the Philosophical Mercury seeing that the philosophical tree requires to be burned by the sun from time to time and then refreshed by water.

It is then necessary to alternate the dry and the damp in order to hasten the process of putrefaction. Once it is noticed that the earth is begin-fling to dessicate, the imbuings are stopped and then one leaves it to dry itself until it reaches a suitable state of dryness and one repeats in this manner until the earth appears like black wax. Then the putrefaction is perfect.

We must now remember what was said during the first procedure in order not to allow the spirit to volatilize or to burn the flowers by stopping the external fire in good time as soon as the putrefaction is total.

The black color that is obtained at the end of forty or fifty days every time that the external fire has been well cared for is a proof that the common gold has been changed into a black earth and this the philosophers call their horse manure.

In the same way as horse manure acts by the power of its proper fire, so, by parallel, our black earth dries its proper oily dampness by means of its own double fire and converts itself after having drunk all its own distilled water and become gray into a white powder which is called air by the philosophers. This constitutes coagulation as we have already described regarding the first procedure.

Once the matter is white, coagulation having ended, it is fixed by taking the matter to a greater state of dessication by the help of the external fire, following the same procedure as in our former coagulation, until the white color is changed into a red color that the philosophers call the element of fire. The matter reaches a very high degree of fixity by itself, to the point that it no longer fears the attacks of the external or ordinary fire which can no longer be prejudicial to it in any way.

Not only is it necessary to fix the matter as we have just demonstrated, but it is also necessary to lapidify it by taking it to the point that it has the appearance of crushed stone by using an ardent fire, that is to say, the first fire used, and following the same means as previously described in order to change the impure part of the matter into a fixed earth and in depriving the matter also of its saline humidity. Then we go on to the separation of the pure from the impure parts of the matter. This is the last degree of regeneration which terminates in the solution.

To arrive at this point, after having well crushed the matter and having it in a subliminatory vase, as we have already said, of three or four fingers in height, thick as ordinary glass, some mercurial water is poured over it. This is our Azoth dissolved in the quantity of Astral Spirit which is necessary to it and as previously indicated by graduating its fire in such a manner as to maintain a temperate heat and giving it, at the last, a quantity of this matter. By this means we transmit all the spiritual part of this last into the water and the earthy part sinks to the bottom. The extract is now decanted and placed in ice in order that the oily quintessence comes together and floats to the top of the water like an oil. The remaining earth is thrown to the bottom as useless, because it was this that held the medicinal virtue of the gold imprisoned which means, that it is of no value.

This oil now floating on the top is separated with the help of the white feather of a pigeon that has been well washed and dampened and it is necessary to be careful not to lose any of it at all because it is the true quintessence of the regenerated common gold in which the three principles are reunited and can no longer be separated the one from the other.

Be careful to note that it is not necessary to push the lapidification of the matter too far in order not to transform the calcined gold into a sort of crystal. It is necessary to control the external fire skillfully so that it dries the saline humidity of the calcined gold, little by little, and thus changes it into a soft earth, which falls like an ash, because of its lapidification or more ample dessication.

The oil thus obtained by separation is the tincture or the sulphur or the radical fire of the gold or the true coloration. It is also the true Potable Gold or the Universal Medicine for all the ills that afflict humanity. One only takes this at the two equinoxes in a quantity necessary lightly to color a soup spoon full of white wine or of distilled dew, seeing that a large quantity of this medicine would destroy the radical humidity of man and thus deprive him of life.

This oil assumes on all possible forms and transforms itself into powder, into salt, into stone, into spirit, etc., by means of dessication by the help of its own secret fire. This oil is also the Blood of the Red Lion.

The ancients symbolized it by the image of a winged dragon resting on the earth. Finally, this oil that cannot be consumed is the Golden Mercury. Once it is made, it is to be divided into two equal parts. One part is kept in the oily state in a small well stoppered flask of white glass. This is kept in a dry spot in order to use it to make the imbuings during the reigns of Mars and the sun as I shall describe at the end of the third procedure. The other portion is dried until it is reduced to a powder by following the same means that I indicated previously to dry the matter and coagulate it. Then this powder is divided into two equal parts. One part is dissolved in proportions of one part to four of its weight of philosophical mercury in order to permeate the other parts of the powder that has been kept back.

THIRD PROCEDURE

CONJUNCTION OF THE SULPHUR WITH
THE MERCURY OF THE PHILOSOPHERS

It is here that all the philosophers begin nearly all their procedures. This has led many people into error. It is also during this procedure that the sulphur of the philosophers is united with their mercury. Nearly all the sages have called this last procedure "fermentation" considering that it is during this that the new sulphur will be dissolved again. It now ferments, it putrefies and is resuscitated by its new regeneration and now with a ten-fold power.

This procedure differs from the two preceding ones. It is for this reason that the philosophers have divided it into seven degrees, to each of which they attribute a planet.

In order to undertake this operation, it is necessary to take half of the powder that has been kept back, of which I already have spoken, and to dampen it little by little in view that the imbuing agent used in too great quantities will again dissolve the oil into a sulphur which will sublimate itself and float over the water. This hinders the union of the sulphur and the mercury, a serious fault that has hindered the success of many philosophers. The matter must therefore be imbued, drop by drop, by sprinkling, in order to bring about the reunion of the moon with the Sun of the Angels and thus forming a thick paste.

The external fire used in order to make these imbuings is that of which we have already spoken when we dissolved a quarter of the golden oil reduced to powder in the necessary quantity of philosophical mercury. This external fire is controlled by the quantity of the matter.

It is now necessary to be careful to maintain the matter in an oily state by reiterated imbuitions as often as it is necessary to make the matter swell and have it start fermenting. Its solution is ended when the matter takes on a bluish color. This solution is called rebis or double mercury and the degree of mercury. This solution is immediately followed by fermentation. Then one stops the dampings and the external fire, allowing the internal fire of the matter work all alone and by itself until the matter has fallen to the bottom of the vase where it becomes as black as coal. It is then that the first degree called that of Saturn is started and that one distills without fire. The liquid that floats above the black matter in the same way described in the two preceding operations.

The black matter is allowed to dry by itself and, as soon as it has reached a suitable state of dryness, one again permeates it with the exterior fire. The permeation is stopped when the matter begins to dry up. It is allowed to acquire a certain degree of dryness by itself and one goes on reiterating in this manner until the matter has reached its total putreficiation. Then the exterior fire is stopped in order not to damage the matter.

By means of the action of the matter's proper fire this now, from being black, becomes gray without our having been obliged to apply the external fire. We have now reached the degree of jupiter. It is in this degree that one notices the colors of the rainbow appear. These are then replaced by a type of browny-black skin which acquires a certain dryness, divides itself, and becomes gray and one notices a small white circle on the walls of the vase.

The matter having reached this point one could use it as a medicine. In that case, it is necessary to allow the matter to become quite dry and transform itself into a white powder by using the same procedures already described in order to obtain this color that will later be made to become red by the aid of the Secret Fire.

This medicine would now have a virtue ten times as great as the first of which I have spoken. But if one desires to use it for the transmutation of metals, after having dried it well, one does not wait for it to become white but makes it such by amalgamating it to equal parts of a common commercial mercury which has been carefully purified by distillation, well sublimated and revivified. This is the milk or the fat of the earth.

In fact, when the common mercury is amalgamated with the matter the whole dissolves itself into a white liquid that looks like milk. It is fixed by the matter into a fixed salt by the action of its proper fire.

Then one again starts the mercurial washings which make it as white as crystal. This is done with the help of seven different washings, during each of which revivified mercury is added in equal parts as I have said above, then by a half, by a third, by a fourth, by a fifth, by a sixth and seventh parts of the weight of the fixed matter so that the weight of the matter will always be greater than that of the revivified mercury being used.

But from the very first washing with equal parts it is essential not to let the fire go out, either by day or night, that is to say the imbuings of the distilled liquid which contains the fire of the matter, so that this latter does not become cold and lost. The compound is the Brass of the Philosophers which must be whitened by frequent dampings until the amalgamated mercury is fixed by our matter, supported by its proper fire. This ends the degree of Jupiter. In continuing thus, the brass takes on a yellowish color. Then it becomes bluish and then the most beautiful white appears on top. Then begins the degree of the Moon. This beautiful white has the appearance of crushed diamonds. It has become a very fine and very subtile powder. One has now obtained the fixed white. A little of this is placed on top of a blade made of reddened copper. If it melts without smoke, that means that the tincture is sufficiently fixed. If the contrary is the case, it is necessary to add fire, continuing this until it has achieved its convenient degree of fixity and there one stops if one only wishes to make the white tincture. A one part of this transmutes one hundred parts of common mercury into a silver better than that of the mines.

But if one wishes to make a red tincture, the fire of the matter must be continued without having allowed it to become cold, that is if one wishes that it become red.

If we continue to apply the external fire, the matter becomes very fine and of such a subtility that is almost unimaginable. For this reason the fire must be very well controlled so that the matter does not volatilize due to the strength of the fire which must penetrate it entirely, but that it stays at the bottom of the vase and transforms itself into a green powder. This is now the degree of Venus.

Wisely continuing the external fire, the matter turns to a lemon yellow. This is the degree of Mars.

This color increases in intensity and becomes the color of copper. When it has reached this point it can no longer increase in intensity by itself. Now it is necessary to have recourse to the golden red mercury, that is to say, to our oil that has been kept and to impregnate the matter with this oil until it has become red. Now begins the degree of the sun.

In continuing the impregnations with the golden oil, the matter becomes redder and redder, then purple and, finally, of a brownish-red. This constitutes the salamander of the wise that fire can no longer attack.

Finally, the matter is conjoined with the same aurific oil by imbuing it, drop to drop, until the oil of the sun is coagulated with the matter and that this latter, placed on a hot blade, melts without smoke. It is by this means that the red tincture is obtained. This is the fixed and coagulating gold of which one part transmutes a hundred parts of mercury into a gold that is better than that of nature.

MULTIPLICATION

The two tinctures of which I have just written, the white and the red, can multiply, both in quality as well as quantity, so long as these tinctures have not been subjected to a common fire which has the effect of making them lose their radical humidity which fixes them into an earth resembling a sutone. In order to achieve the multiplication of the white and red tinctures, the third process has to be repeated in its entirety.

The white and red powders have to be dissolved in the philosophical mercury. They must undergo fermentation and putrefaction as well as regeneration. In order to achieve success, the imbitions must be reiterated little by little. The fire must be directed and controlled in the manner already described. After this second multiplication one part is enough to make a projection on a thousand parts of mercury and transmute it into silver or into gold, depending on the color of the powder. These metals are perfect.

The multiplication in quality is obtained by reiterating the philosophical sublimation which takes place when the pure is separated from the impure with the help of the Philosophical Mercury. The manipulations of the third process are to be punctiliously repeated after having dried and reduced into powder all the white oil with the help of the fire of the matter -- if working in the white -- and

one part of the red oil if working in the red. This is
in order to keep the other part to be used in the
degree of Mars and the Sun, as I have already stated,
if operating with the red.

The multiplication in quantity is achieved by
adding regenerated common mercury as already stated.
If one wishes to make the multiplication in quality at
the same time, it is necessary to begin, as a general
rule, by sublimating the matter. The pure must be
separated from the impure, by total descication -- if
working with the white -- or by half if working with
the red. This by the help of the self-same fire which
is to be controlled in the same manner as I did for
the third process. This is to reduce them into a
powder which we shall divide each into equal parts.
One part will be dissolved in four parts its weight in
philosophical mercury. This will serve to imbue the
part that has been retained by absolutely reiterating
the third operation.

Should one so wish, these manipulations may be
reiterated up to ten times. Each time the matter will
acquire a tenfold power and will become so subtile
that the last time, it will pass through the glass,
becoming totally volatilized. Usually one stops at the
ninth multiplication. At this point it becomes so
volatile that at the least heat it pierces the glass
and evaporates. As a result, it is usual to stop at
the transmutation of one part in a thousand or ten
thousand at the most, in order not to risk losing so

precious a treasure.

At this point, I shall not describe the very strange processes I have done, to my great astonishment, both in the vegetable and the animal kingdoms as well as the means used for making malleable glass, pearls and precious stones more beautiful even than those of nature, following the procedure indicated by Zachary in using vinegar and the matter fixed in white, and grains of pearls and rubies fine powdered moulded and then fixed by means of the fire of the matter. I restrain my words in order not to perjure myself and seem here to fly beyond the boundaries of human credulity.

Having finished my work, I took one hundred grammes of distilled mercury and placed it in a crucible. As soon as they began to smoke, I threw one gramme of my transmutating sulphur on top. It turned into an oil on top of the mercury and I saw this last successively coagulating more and more. I then increased my fire and thus pursuing made it stronger to the point that my mercury became perfectly fixed. This took about an hour. Having poured it into a little ingot—mould, I tested it and found it better than that of the mines.

How great was my joy! I was carried away with rapture! Like Pygmalion, I fell upon my knees in order to contemplate my work and to thank the Eternal. I wept a torrent of tears! How sweet they were. How my heart was relieved! It would be hard for me here to describe everything I felt and the position in which I found myself. A thousand thoughts came simultaneously before my mind. The first would lead me to seek the King citizen in order to confess my triumph; another, that one day I would make enough gold to create various foundations in the town that had given me birth. Another thought made me want to see married as many young girls as there are sections in Paris, dowering each one. Another idea impelled me to find out the addresses of people poor and ashamed of their poverty and to go in person to their homes to bring them help. In the end, I began to fear that joy had cracked my wits. I experienced the need to use violence on myself and to take a great deal of exercise by walking in the country. I did this during eight consecutive days. Every few hours I would take off my hat and raise my eyes to the sky to thank heaven for having granted me such a blessing. The tears would flow abundantly from my eyes. Finally, I managed to calm myself and to realize how I would expose myself to danger if I undertook such endeavors. After having reflected soberly, I came to the firm decision that I would live unknown, without pomp, and restrain my ambition by giving happiness in secret without letting myself be known as a benefactor.

I had told my wife of my success and promised to repeat the transmutation for her benefit. She urged me not to speak of this to anyone. It was Maundy Thursday 1831 at seven minutes past ten in the morning that I

236

had achieved my first transmutation, alone. I lacked mercury and put off showing my wife until the day after Easter. From a gardener, I bought a laurel branch and a twig of evergreen. Having tied them together, I wrapped the whole in a sheet of writing paper and wended my way home where my wife was sitting by the window, reading. Kneeling in front of her, I placed my bouquet at her feet, saying: "Here it is at last, dear friend, placed at your feet. It crowns me at last, just as you and I are descending towards the grave. It has cost me thirty-seven years of painful labors and more than fifteen hundred nights without sleep. I have known so many humiliations, been overwhelmed with abuses, shunned by my friends, rejected by your family and mine. Finally, I lost the most interesting creature that one might possibly see. Yet, I have never ceased from being an honest man and cherishing you." My head fell upon her knees. I began to weep. 0 tears of regret at remembering my losses! Tears at the tribulations that I have known! Tears of joy, how sweet you were! You calmed my heart! I was reborn. I was a new man. Her eyes filling with tears, my wife lifted my head, saying: "Stand up, my friend, and stop crying." I placed my lips upon hers and this tenderly reciprocated kiss embellished the sweetness of my life and reanimated my mind beaten down by unhappiness.

It was not enough to have admitted my success and to have placed the laurel at her feet. It was necessary to convince her and to perform the transmutation in front of her.

I took a watchglass and put therein a small quantity of just bought distilled commercial mercury which was quite pure. On this I placed my transmutatory mercury in the oily state, not powdered, in the proportion of one part to a hundred. I shook the glass gently in such a way as to give the oil a circulatory movement. Joy fully we saw the mercury offer a most curious phenomenon. It coagulated into the color of the most beautiful gold. All I had to do was to melt it in a crucible and pour it off. Thus, to my wife's great astonishment, I demonstrated a cold transmutation. She then said to me: "Your success is the crown of your desires. If you wish to make me happy and help me forget the long chain of our miseries, let us spend the rest of our days unknown, without making any sort of show. Hide everything that might reveal your secret and serve as a lure to the greed of those whose ambition is insatiable. Hide from intrigue, from tyranny, from base humanity."

I replied: "I have sworn that, even if melted lead were to be poured through my veins, I would take my secret with me to the grave. That is to say, the knowledge of the Matter, the Fire, and the Labors of Hercules. Before God I swear to you that I shall make you happy and fulfill all your desires. Let us hope that the Eternal will protect us from men who are filled with envy, from the vicious and the corrupt."

You young people who are likely to read this work, do not let your wish to make a showing in the world and the lure of wealth urge you enter into the search for the Philosopher's Stone. If you could only know, as I do, the various hardships that I have undergone in order to reach this goal, you would draw back in fear. Start only if God allows you to meet a man who has already succeeded in making the Stone. One who will lead you by the hand from the beginning to the end. Push away from your mind in horror the idea of devoting yourself to Hermetic Philosophy. Whole secrets are unbelievably far more difficult to find on one's own. If, hoping to have better luck than I did, you reject my advice and are so fortunate as to succeed, never forget those more unfortunate than your-self. Above all, be discreet. Be miserly in what you spend to satisfy your tastes and your passions, but be lavish towards the poor. Remember that the greatest happiness for a well-born heart is to make others happy without their knowledge and always keep the Eternal before your mind.

Flee the corrupt of society. They have all the means to abuse your qualities. They would seem ruined by their promises that seem to flow from a good heart, but they become rich by making you their dupe. In few words, do not seek happiness in life from the two extremes of society. Seek it in the middle class, from honest industrialists. However, there are certain exceptions to be made and I would be an ingrate to judge otherwise. Never in my life shall I forget a well-born man to whom I have promised myself I shall give proof of my friendship.

Estimable youth, may my life serve you as an example and my suggestions as lessons, and may they merit a few tears from your eyes to sweeten the long chain of miseries I have known.

If you, Kings of the earth, knew the number of people who devote their lives to seeking the Philosopher's Stone, you would be astonished. If, also, you knew that during three or four hundred years only one or two men succeed in obtaining that which, in business, does not offer the product of a single gold mine in Peru or elsewhere. Instead of seeking out those who have succeeded and torturing them, you would cover them with goodness, grant them your friendship and help so that they could fully serve suffering humanity and allow you to share in the benefits of their discoveries.

My fellow countrymen, if the Eternal allows me to leave you that which my heart wishes to be your fate, in gratitude do the following for me.

Transport my mortal remains to a chalky place facing a little turret on which lies a sorrowful emblem from an ancient war, at the base of which runs a rivulet whose source is a league from there and turns several mills. Have them covered only by a large block of hard granite which is common in the place where I was born and where I was married and on it put

240

just the following inscription:

"HERE REST THE MORTAL REMAINS OF THE
   UNFORTUNATE CYLIANI."

Seeing that no law exists in any country forbidding the publishing of a discovery useful to society, I have had this work printed. I have also put into circulation gold, perfect by its weight, its color, its specific weight and fusibility. By what right would one wish to give preference to gold that is mined over that created by the philosopher's art, considering that this latter is the better.

FINIS.

A Word from the Publisher

Thank you for purchasing this small work from The R.A.M.S. Library of Alchemy. During his lifetime, Hans Nintzel was dedicated to the identification, acquisition, study, retyping and, when necessary, translation of what he considered to be the most important known works on Alchemy. Hans was assisted by his sparse network of fellow Alchemists, all members of the Restorers of Alchemical Manuscripts Society (R.A.M.S.). I was an active member of R.A.M.S.

My goal is to publish all of the works originally made available through R.A.M.S. as photocopies. To facilitate this, I have chosen to have the books professionally printed. I also have a few titles that I intend to add to the original R.A.M.S. Library, selected by strict criteria established by Hans.

If you have a work on Alchemy that you believe should be a part of the R.A.M.S. Library, please contact me through R.A.M.S. Publishing Company.

Philip N. Wheeler

www.ingramcontent.com/pod-product-compliance
Lightning Source LLC
Chambersburg PA
CBHW080803180526
45168CB00006B/2312